Bad to the Bone

Crafting Electronic Systems with BeagleBone and BeagleBone Black

Synthesis Lectures on Digital Circuits and Systems

Editor
Mitchell A. Thornton, *Southern Methodist University*

The Synthesis Lectures on Digital Circuits and Systems series is comprised of 50- to 100-page books targeted for audience members with a wide-ranging background. The Lectures include topics that are of interest to students, professionals, and researchers in the area of design and analysis of digital circuits and systems. Each Lecture is self-contained and focuses on the background information required to understand the subject matter and practical case studies that illustrate applications. The format of a Lecture is structured such that each will be devoted to a specific topic in digital circuits and systems rather than a larger overview of several topics such as that found in a comprehensive handbook. The Lectures cover both well-established areas as well as newly developed or emerging material in digital circuits and systems design and analysis.

Bad to the Bone: Crafting Electronic Systems with BeagleBone and BeagleBone Black
Steven Barrett and Jason Kridner
2013

Introduction to Noise-Resilient Computing
S.N. Yanushkevich, S. Kasai, G. Tangim, A.H. Tran, T. Mohamed, and V.P. Smerko
2013

Atmel AVR Microcontroller Primer: Programming and Interfacing, Second Edition
Steven F. Barrett and Daniel J. Pack
2012

Representation of Multiple-Valued Logic Functions
Radomir S. Stankovic, Jaakko T. Astola, and Claudio Moraga
2012

Arduino Microcontroller: Processing for Everyone! Second Edition
Steven F. Barrett
2012

Advanced Circuit Simulation Using Multisim Workbench
David Báez-López, Félix E. Guerrero-Castro, and Ofelia Delfina Cervantes-Villagómez
2012

Embedded Systems Design with the Atmel AVR Microcontroller: Part I
Steven F. Barrett
2009

Embedded Systems Interfacing for Engineers using the Freescale HCS08 Microcontroller II:
Digital and Analog Hardware Interfacing
Douglas H. Summerville
2009

Designing Asynchronous Circuits using NULL Convention Logic (NCL)
Scott C. Smith and JiaDi
2009

Embedded Systems Interfacing for Engineers using the Freescale HCS08 Microcontroller I:
Assembly Language Programming
Douglas H.Summerville
2009

Developing Embedded Software using DaVinci & OMAP Technology
B.I. (Raj) Pawate
2009

Mismatch and Noise in Modern IC Processes
Andrew Marshall
2009

Asynchronous Sequential Machine Design and Analysis: A Comprehensive Development of
the Design and Analysis of Clock-Independent State Machines and Systems
Richard F. Tinder
2009

An Introduction to Logic Circuit Testing
Parag K. Lala
2008

Pragmatic Power
William J. Eccles
2008

Multiple Valued Logic: Concepts and Representations
D. Michael Miller and Mitchell A. Thornton
2007

Finite State Machine Datapath Design, Optimization, and Implementation
Justin Davis and Robert Reese
2007

Atmel AVR Microcontroller Primer: Programming and Interfacing
Steven F. Barrett and Daniel J. Pack
2007

Pragmatic Logic
William J. Eccles
2007

PSpice for Filters and Transmission Lines
Paul Tobin
2007

PSpice for Digital Signal Processing
Paul Tobin
2007

PSpice for Analog Communications Engineering
Paul Tobin
2007

PSpice for Digital Communications Engineering
Paul Tobin
2007

PSpice for Circuit Theory and Electronic Devices
Paul Tobin
2007

Pragmatic Circuits: DC and Time Domain
William J. Eccles
2006

Pragmatic Circuits: Frequency Domain
William J. Eccles
2006

Pragmatic Circuits: Signals and Filters
William J. Eccles
2006

High-Speed Digital System Design
Justin Davis
2006

Introduction to Logic Synthesis using Verilog HDL
Robert B.Reese and Mitchell A.Thornton
2006

Microcontrollers Fundamentals for Engineers and Scientists
Steven F. Barrett and Daniel J. Pack
2006

Bad to the Bone: Crafting Electronic Systems with BeagleBone and BeagleBone Black

Steven Barrett and Jason Kridner

www.morganclaypool.com

ISBN: 9781627051378   paperback
ISBN: 9781627051385   ebook

DOI 10.2200/S00500ED1V01Y201304DCS041

A Publication in the Morgan & Claypool Publishers series
SYNTHESIS LECTURES ON DIGITAL CIRCUITS AND SYSTEMS

Lecture #41
Series Editor: Mitchell A. Thornton, *Southern Methodist University*
Series ISSN
Synthesis Lectures on Digital Circuits and Systems
Print 1932-3166   Electronic 1932-3174

Bad to the Bone

Crafting Electronic Systems with BeagleBone and BeagleBone Black

Steven Barrett
University of Wyoming

Jason Kridner
Texas Instruments

SYNTHESIS LECTURES ON DIGITAL CIRCUITS AND SYSTEMS #41

MORGAN & CLAYPOOL PUBLISHERS

ABSTRACT

BeagleBone is a low cost, open hardware, expandable computer first introduced in November 2011 by BeagleBoard.org, a community of developers sponsored by Texas Instruments. Various BeagleBone variants, including the original BeagleBone and the new BeagleBone Black, host a powerful 32–bit, super–scalar ARM Cortex A8 processor operating from 720 MHz to 1 GHz. Yet, BeagleBone is small enough to fit in a small mint tin box. The "Bone" may be used in a wide variety of projects from middle school science fair projects to senior design projects to first prototypes of very complex systems. Novice users may access the power of the Bone through the user–friendly Bonescript environment, a browser–based experience, in MS Windows, the Mac OS X, or the Linux operating systems. Seasoned users may take full advantage of the Bone's power using the underlying Linux––based operating system, a host of feature extension boards (Capes) and a wide variety of Linux community open source libraries. This book provides an introduction to this powerful computer and has been designed for a wide variety of users including the first time novice through the seasoned embedded system design professional. The book contains background theory on system operation coupled with many well–documented, illustrative examples. Examples for novice users are centered on motivational, fun robot projects while advanced projects follow the theme of assistive technology and image processing applications.

KEYWORDS

BeagleBone, Linux, Ångström distribution, microcontroller interfacing, embedded systems design, Bonescript, ARM, open source computing

For the students!

Contents

Preface

BeagleBone is a low cost, open hardware expandable computer first introduced in November 2011 by BeagleBoard.org, a community of developers sponsored by Texas Instruments. Various BeagleBone variants host a powerful 32–bit, super–scalar ARM Cortex A8 processor operating from 720 MHz to 1 GHz. Yet, BeagleBone is small enough to fit in a small mint tin box. The "Bone" may be used in a wide variety of projects from middle school science fair projects to senior design projects to first prototypes of very complex systems. Novice users may access the power of the Bone through the user––friendly browser–based Bonescript environment that may be operated in MS Windows, the Mac OS X, or the Linux operating systems. Seasoned users may take full advantage of the Bone's power using the underlying Linux–based operating system, a host of feature extension boards (Capes) and a wide variety of Linux community open source libraries. This book provides an introduction to this powerful computer and has been designed for a wide variety of users including the first time novice through the seasoned embedded system design professional. The book contains background theory on system operation coupled with many well–documented, illustrative examples. Examples for novice users are centered on motivational, fun robot projects while advanced projects follow the theme of assistive technology and image processing applications.

Texas Instruments has a long history of educating the next generation of STEM (Science, Technology, Engineering and Mathematics) professionals. BeagleBone is the latest educational innovation in a long legacy which includes the Speak & Spell and Texas Instruments handheld calculators. The goal of the BeagleBone project is to place a powerful, expandable computer in the hands of young innovators. Novice users can unleash the power of the Bone through the user–friendly, browser–based Bonescript environment. As the knowledge and skill of the user develops and matures, BeagleBone provides more sophisticated interfaces including a complement of C–based functions to access the hardware systems aboard the ARM Cortex A8 processor. These features will prove useful for college––level design projects including capstone design projects. The full power of the processor may be unleashed using the underlying onboard Linux–based operating system. A wide variety of hardware extension features are also available using daughter card "Capes" and Linux community open source software libraries. These features allow for the rapid prototyping of complex, expandable embedded systems.

A BRIEF BEAGLE HISTORY

The Beagle family of computer products include BeagleBoard, BeagleBoard xM, the original BeagleBone and the new BeagleBone Black. All have been designed by Gerald Coley, a long time expert Hardware Applications Engineer, of Texas Instruments. He has been primarily responsible for all hardware design, manufacturing planning, and scheduling for the product line. The goal of the entire product line is to provide users a powerful computer at low cost with no restrictions. As Gerald describes it, his role has been to "get the computers out there and then get out of the way" to allow users to do whatever they want. The computers are intended for everyone and the goal is to provide "out of the box success." He indicated it is also important to provide users a full suite of open source hardware details and to continually improve the product line by listening to customers' desires and keeping the line fresh with new innovations. In addition to the computers, a full line of Capes to extend processor features is available. There are more than 20 different Capes currently available with 50 more on the way.

BeagleBone was first conceptualized in early Fall 2011. The goal was to provide a full–featured, expandable computer employing the Sitara ARM processor in a small profile case. Jason Kridner told Gerald the goal was to design a processor board that would fit in a small mint tin box. After multiple board revisions, Gerald was able to meet the desired form factor. BeagleBone enjoyed a fast development cycle. From a blank sheet of paper as a starting point, to full production was accomplished, in 90 days.

New designs at Texas Instruments are given a code name during product development. During the early development of the Beagle line, the movie "Underdog" was showing in local theaters. Also, Gerald had a strong affinity to Beagles based on his long time friend and pet "Jake." Gerald dubbed the new project "Beagle" with the intent of changing the name later. However, the name stuck and the rest is history. Recently, an acronym was developed to highlight the Beagle features:

- **B**ring your own peripherals

- **E**ntry level costs

- **A**RM Cortex–A8 superscalar processor

- **G**raphics accelerated

- **L**inux and open source community

- **E**nvironment for innovators

APPROACH OF THE BOOK

Concepts will be introduced throughout the book with the underlying theory of operation, related concepts and many illustrative examples. Concepts early in the book will be illustrated using moti-vational robot examples including an autonomous robot navigating about a two–dimensional maze based on the Graymark Blinky 602A robot platform, an autonomous four wheel drive (4WD) robot (DFROBOT ROBOT0003) in a three–dimensional mountain maze, and a SeaPerch Remotely Operated Vehicle (ROV).

Advanced examples follow a common theme of assistive technology. Assistive technology provides those with challenges the ability to meet their potential using assistive devices. One example covered in the book is a BeagleBone based Speak & Spell. Speak and Spell is an educational toy developed by Texas Instruments in the mid–1970's. It was developed by the engineering team of Gene Franz, Richard Wiggins, Paul Breedlove and Larry Branntingham.

This book is organized as follows. Chapter 1 provides an introduction to the BeagleBone processor and the Bonescript programming environment. Chapter 2 provides a brief introduction to programming and provides a detailed introduction to the Bonescript environment. The chapter concludes with an extended example of the Blink 602A robot project. Chapter 3 provides electrical interfacing fundamentals to properly connect BeagleBone to peripheral components. Chapter 4 provides an introduction to system level design, the tools used to develop a system and two advanced robot system examples.

Chapter 5 begins an advanced treatment of BeagleBone and the ARM Cortex A8 processor and an in depth review of exposed functions and how to access them using both Bonescript and C functions. In Chapter 6 the full power of BeagleBone is unleashed via several system level examples. Chapter 7 briefly reviews additional resources available to the BeagleBone user.

A LITTLE BEAGLE LORE

The beagle dog has been around for centuries and was quite popular in Europe. They were known for their sense of smell and were used to track hare. They are also known for their musical, bugling voice. In "The Beagle Handbook," Dan Rice, Doctor of Veterinary Medicine (DVM), indicates the Beagle was probably named for their size or voice. A combination of the French words for "open wide" and "throat" results in Begueule and might refer to the Beagle's distinctive and jubilant bugle while on the hunt. Rice also notes "the word beagling isn't found in Webster's dictionary, but it's well recognized in canine fancy. Used by the Beagle community for practically all endeavors that involve their beautiful dogs [Rice, 2000]." We too will use the term "beagling" to describe the fun sessions of working with BeagleBone that are ahead. It is also worth mentioning that Ian Dunbar, Member of the Royal College of Veterinary Surgeons (MRCVS) in "The Essential Beagle" indicates the "overall temperament of the Beagle is bold and friendly." This seems like an appropriate description for the BeagleBone computer running Bonescript software.

THE BEAGLEBOARD.ORG COMMUNITY

The BeagleBoard.org community has many members. What we all have in common is the desire to put processing power in the hands of the next generation of users. BeagleBoard.org, with Texas Instruments' support, embraced the open source concept with the development and release of BeagleBone in late 2011. Their support will insure the BeagleBone project will be sustainable. BeagleBoard.org partnered with circuitco (www.circuitco.com) to produce BeagleBone and its associated Capes. The majority of the Capes have been designed and fabricated by circuitco. Clint Cooley, President of circuitco, is most interested in helping users develop and produce their own ideas for BeagleBone Capes. Texas Instruments has also supported the BeagleBoard.org community by giving Jason Kridner the latitude to serve as the open platform technologist and evangelist for the BeagleBoard.org community. The most important members of the community are the BeagleBoard and Bone users. Our ultimate goal is for the entire community to openly share their successes and to encourage the next generation of STEM practitioners.

A LITTLE BIT ABOUT THE AUTHORS

Steve is a life time teacher. He has taught at a variety of age levels from middle school science enhancement programs through graduate level coursework. He served in the United States Air Force for 20 years and spent approximately half of that time as a faculty member at the United States Air Force Academy. Following military "retirement," he began a second academic career at the University of Wyoming as an Assistant Professor. He now serves as a Professor of Electrical and Computer Engineering and the Associate Dean for Academic Programs. He is planning on teaching into his 80's and considers himself a student first teacher. Most importantly, he has two "grand beagles," Rory and Romper, fondly referred to as "the girls."

Jason got an early start with computing at age 9 programming his mom's Tandy Radio Shack TRS–80. He was also a big fan of Forrest Mim's "Getting Started in Electronics." Much of his allowance was spent developing projects. He really enjoyed the adventure of trying new hardware and software projects. His goal is to bring back this spirit of adventure and discovery to the Beagle-Board.org community. While still in high school, he worked extensively with AutoCAD at a leak and flow testing company. He joined Texas Instruments in 1992 after a co–op with them while a student at Texas A&M University. He started using Linux at about the same time. Since joining TI he has worked on a wide variety of projects including audio digital signal processing, modems, home theater sound, multi–dimensional audio and MP3 player development.

Steven Barrett and Jason Kridner
April 2013

Acknowledgments

The authors would like to thank Cathy Wicks and Larissa Swanland of Texas Instruments who proposed this collaboration. We also thank Gerald Coley of Texas Instruments who was interviewed for the book. We also thank Clint Cooley, President of circuitco for hosting a tour of his company where BeagleBone and its associated Capes are produced. We would also like to thank Joel Claypool of Morgan & Claypool Publishers for his support of this project and his permission to use selected portions from previous M&C projects. We would also like to thank our reviewers including Dimple Joseph a student in Houston, Texas; Richard (Andy) Darter and Luke Kaufman of the University of Wyoming; Dr. Derek Molloy of Dublin City University; Shawn Trail of the University of Victoria, Victoria, BC, Canada; and C. P. Ravikumar, Technical Director of University Relations at TI India. Also, a special thank you to Jonathan Barrett of Closer to the Sun International

Steven Barrett and Jason Kridner
April 2013

CHAPTER 1

Getting Started

Objectives: After reading this chapter, the reader should be able to do the following:

- Provide a brief history of the Beagle computer line.

- Outline the different members of the BeagleBoard.org community.

- Appreciate the role of the BeagleBoard.org community.

- Describe BeagleBone concept of open source hardware.

- Diagram the layout of the original BeagleBone and BeagleBone Black computers.

- Summarize the differences between the original BeagleBone and BeagleBone Black computers.

- Describe BeagleBone Cape concept and available Capes.

- Define the power requirements for BeagleBone computer.

- Download, configure, and successfully execute a test program using the Cloud9 Integrated Development Environment (IDE) and the Bonescript software.

- Design and implement a BeagleBone Boneyard prototype area to conduct laboratory exercises.

1.1 WELCOME!

Welcome to the wonderful world of BeagleBone! Whether you are a first–time BeagleBone user or are seasoned at "Beagling," this book should prove useful. Chapter 1 is an introduction to BeagleBone, its environment and the Beagle community. BeagleBone hosts the Linux operating system; however, the user–friendly Bonescript programming environment may be used in a wide variety of browser environments including: Microsoft Windows, MAC OS X, and Linux. We provide instructions on how to rapidly get up–and–operating right out of the box! Also, an overview of BeagleBone features and subsystems is provided. The chapter concludes with several examples to get you started.

1.2 OVERVIEW

BeagleBone is a low cost, open hardware, expandable computer first introduced in November 2011 by BeagleBoard.org, a community of developers started by Beagle enthusiasts at Texas Instruments. Various BeagleBone variants host a powerful 32–bit, super–scalar ARM® Cortex^{TM} –A8 processor

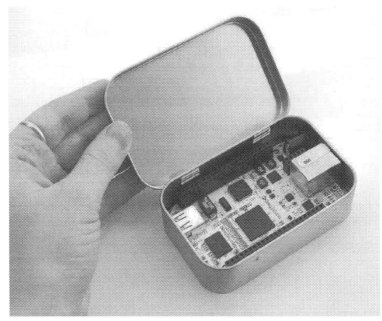

Figure 1.1: BeagleBone. (Figures adapted and used with permission of `www.adafruit.com`.)

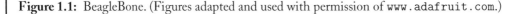

operating up to 1 GHz. This allows the "Bone" to be used in a wide variety of applications usually reserved for powerful, desktop systems. The Bone is a full featured computer small enough to fit in a small mint tin box as shown in Figure 1.1. The combination of computing power and small form factor allows the Bone to be used in a wide variety of projects from middle school science fair projects to senior design projects to first prototypes of very complex systems.

Novice users may access the power of the Bone through the user–friendly, browser–based Bonescript environment in MS Windows, Mac OS X, and Linux. Seasoned users may take full advantage of the Bone's power using the underlying Linux–based operating system, a host of feature extension boards (Capes) and a wide variety of open source libraries.

Texas Instruments has supported BeagleBone development and has a long history of educating the next generation of STEM (Science, Technology, Engineering and Mathematics) professionals. BeagleBone is the latest education innovation in a long legacy which includes the Speak & Spell and Texas Instruments handheld calculators. The goal of BeagleBone project is to place a powerful, expandable computer in the hands of young innovators through the user–friendly, browser–based Bonescript environment. As the knowledge and skill of the user develops and matures, BeagleBone provides more sophisticated interfaces including a complement of C–based functions to access the hardware systems aboard the ARM Cortex A8 processor. These features will prove useful for college-

–level design projects including capstone design projects. The full power of the processor may be unleashed using the underlying onboard Linux–based operating system.

To assemble a custom system, the Bone may be coupled with a wide variety of daughter board "Capes" and open source libraries. These features allow for the rapid prototyping of complex, expandable embedded systems. A full line of Capes to extend processor features is available. There are over 35 different Capes currently available with at least as many more on the way. The Cape system is discussed later in this chapter. Open source libraries are discussed later in the book.

1.3 A BRIEF BEAGLE HISTORY

The Beagle family of computer products include BeagleBoard, BeagleBoard–xM, the original BeagleBone and the new BeagleBone Black. All have been designed by Gerald Coley, a long time expert Hardware Applications Engineer, of Texas Instruments. He has been primarily responsible for all hardware design, manufacturing planning, and scheduling for the product lines. The goal of the entire platform line is to provide users a powerful computer at a low cost with no restrictions. As Gerald describes it, his role has been to "get the computers out there and then get out of the way" to allow users to do whatever they want. The computers are intended for everyone and the goal is to provide out–of–box success." He indicated it is also important to provide users a full suite of open source hardware details and to continually improve the product line by listening to customers' desires and keeping the line fresh with new innovations.

BeagleBone was first conceptualized in early Fall 2011. The goal was to provide a full–featured, expandable computer employing the Sitara^{TM} AM335x ARM$^{®}$ processor in a small profile case. Jason Kridner, a longtime expert Software Applications Engineer responsible for all Beagle software and co–author of this textbook, of Texas Instruments told Gerald the goal was to design a processor board that would fit in a small mint tin box. After multiple board revisions, Gerald was able to meet the desired form factor. BeagleBone enjoyed a fast development cycle. From a blank sheet of paper starting point to full production was accomplished in 90 days.

New designs at Texas Instruments are given a code name during product development. During the early development of the Beagle line, the movie "Underdog" was showing in local theaters. Also, Gerald had a strong affinity to Beagles based on his long time friend and pet, "Jake." Gerald dubbed the new project "Beagle" with the intent of changing the name later. However, the name stuck and the rest is history.

The "Underdog" movie was adapted from the cartoon series of the same name. The series debuted in 1964 and featured the hero "Underdog" –who was humble and loveable "Shoeshine Boy" until danger called. Shoeshine Boy was then transformed into the invincible flying Underdog complete with a cape that allowed him to fly. His famous catch phrase was "Have no fear, Underdog is here!" This series is readily available on Amazon and is a lot of good fun. BeagleBone is a bit like Underdog. It can fit into an unassuming, mint tin box, yet is a full–featured, "fire–breathing," computer equipped with the Linux operating system whose features are extended with Capes.

1.4   BEAGLEBOARD.ORG COMMUNITY

The BeagleBoard.org community has thousands of members. What we all have in common is the desire to put processing power in the hands of the next generation of users. BeagleBoard.org with Texas Instruments support embraced the open source concept with the development and release of BeagleBone in late 2011. Their support will ensure BeagleBone project remains sustained. Beagle-Board.org partnered with Circuitco (a contract manufacturer) to produce BeagleBone and many of its associated Capes. The majority of the Capes have been designed and fabricated by Circuitco. Clint Cooley, President of Circuitco, is most interested in helping users develop and produce their own ideas for BeagleBone Capes. Texas Instruments has also supported the BeagleBoard.org community by giving Jason Kridner the latitude to serve as the open platform technologist and evangelist for the BeagleBoard.org community. The most important members of the community are the BeagleBoard and BeagleBone users (you). Our ultimate goal is for the entire community to openly share their successes and to encourage the next generation of STEM practitioners.

In the next several sections, an overview of BeagleBone hardware, system configuration and software is discussed.

1.5   BEAGLEBONE HARDWARE

There are two different variants of BeagleBone: the original BeagleBone, released in late 2011, and BeagleBone Black released, in early 2013. The features of both Bones are summarized in Figure 1.2.

The two variants of BeagleBone are physically quite similar. They both fit inside a small mint box. The original BeagleBone is equipped with a Sitara^{TM} AM3359 ARM$^{®}$ processor that operates at a maximum frequency of 720 MHz when powered from an external 5 VDC source. The original BeagleBone is also equipped with an SD/MMC micro SD card. The card acts as the "hard drive" for BeagleBone board and hosts the Ångström distribution operating system, Cloud9 Integrated Development Environment (IDE), and Bonescript. The original BeagleBone also features an integrated JTAG "and serial port" over the USB client connection [Coley].

BeagleBone Black hosts the Sitara^{TM} AM3358 ARM$^{®}$ processor which operates at a maximum frequency of 1 GHz. The micro SD card provided with the original BeagleBone has been replaced with a 2 Giga byte (GB) eMMC (embedded MultiMedia Card) on BeagleBone Black. This provides for non–volatile mass storage in an integrated circuit package. The eMMC acts as the "hard drive" for BeagleBone Black board and hosts the Ångström distribution operating system, Cloud9 Integrated Development Environment (IDE) and Bonescript. The micro SD card connector is still available for expanding the storage. BeagleBone Black has also been enhanced with an HDMI (High–Definition Multimedia Interface) framer and micro connector. HDMI is a compact digital audio/interface which is commonly used in commercial entertainment products such as DVD players, video game consoles and mobile phones. Interestingly, BeagleBone Black costs approximately half of the original BeagleBone at approximately US $45 [Coley].

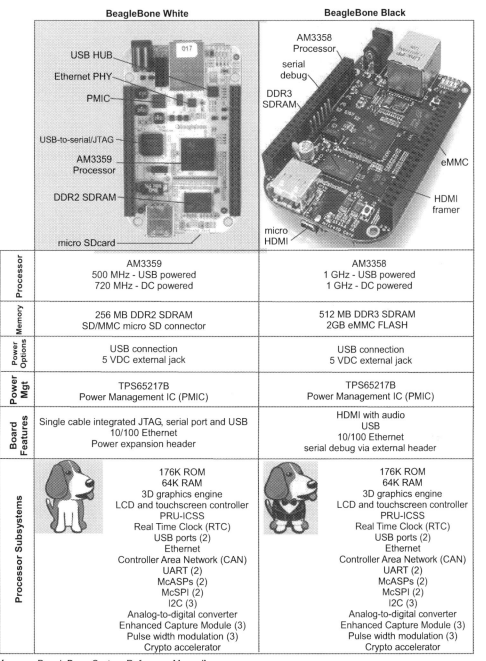

BeagleBone White	BeagleBone Black
Processor	
AM3359 500 MHz - USB powered 720 MHz - DC powered	AM3358 1 GHz - USB powered 1 GHz - DC powered
Memory	
256 MB DDR2 SDRAM SD/MMC micro SD connector	512 MB DDR3 SDRAM 2GB eMMC FLASH
Power Options	
USB connection 5 VDC external jack	USB connection 5 VDC external jack
Power Mgt	
TPS65217B Power Management IC (PMIC)	TPS65217B Power Management IC (PMIC)
Board Features	
Single cable integrated JTAG, serial port and USB 10/100 Ethernet Power expansion header	HDMI with audio USB 10/100 Ethernet serial debug via external header
Processor Subsystems	
176K ROM 64K RAM 3D graphics engine LCD and touchscreen controller PRU-ICSS Real Time Clock (RTC) USB ports (2) Ethernet Controller Area Network (CAN) UART (2) McASPs (2) McSPI (2) I2C (3) Analog-to-digital converter Enhanced Capture Module (3) Pulse width modulation (3) Crypto accelerator	176K ROM 64K RAM 3D graphics engine LCD and touchscreen controller PRU-ICSS Real Time Clock (RTC) USB ports (2) Ethernet Controller Area Network (CAN) UART (2) McASPs (2) McSPI (2) I2C (3) Analog-to-digital converter Enhanced Capture Module (3) Pulse width modulation (3) Crypto accelerator

[source: BeagleBone System Reference Manual]

Figure 1.2: BeagleBone comparison [Coley].(Figures adapted and used with permission of beagle-board.org.)

1.5.1   OPEN SOURCE HARDWARE

To spur the development and sharing of new developments and features among the BeagleBoard community, all hardware details of BeagleBone boards are readily available as open source. This includes detailed schematics, bill–of–materials and printed circuit board (PCB) layout. This allows for custom designs to take full advantage of BeagleBone–based products at the lowest possible cost. Hardware details are readily available at `www.beagleboard.org`.

1.6   DEVELOPING WITH BONESCRIPT

Bonescript helps to provide a user–friendly, browser–based programming environment for Beagle-Bone. Bonescript consists of a JavaScript library of functions to rapidly develop a variety of physical computing applications. Bonescript is accessed via a built–in web server that is pre–installed on the BeagleBone board. Because Bonescript programming is accessed over your browser, it may be developed from any of MS Windows, Mac OS X, and Linux. Exercise 1 provided later in the chapter provides easy–to–use instructions on how to quickly get Bonescript-based program up and operating. Although Bonescript is user–friendly, it may also be used to rapidly prototype complex embedded systems.

1.7   BEAGLEBONE CAPES

A wide variety of peripheral components may be easily connected to BeagleBone. A series of Capes have been developed to give BeagleBone "super powers" (actually, using these plug–in boards provide BeagleBone with additional functionality). Currently, there are over 20 different capes available with many more on the way. See `www.beaglebonecapes.com` for the latest information on available capes. BeagleBone may be equipped with up to four Capes at once. Each Cape is equipped with an onboard EEPROM to store Cape configuration data. For example, Capes are available to provide BeagleBone:

- a seven inch, 800 x 400 pixel TFT LCD (liquid crystal display) (LCD7 Cape)

- a battery pack (Battery Cape)

- a Controller Area Network (CAN) interface (CANbus Cape)

- separate Capes for interface to a wide variety of devices including DVI–D, Profibus, RS–232, RS–485 and VGA standards

- an interface for two stepper motors

There is also a Cape equipped with a solderless breadboard area to provide a straight forward interface to external components. This Cape is illustrated in Figure 1.3 along with a representative sample of other Capes.

Figure 1.3: BeagleBone Capes. (top left) prototype Cape, (top right) Liquid Crystal Display LCD7 Cape, (bottom left) Controller Area Network Cape, (bottom right) motor control Cape [Circuitco].

Most of these Capes have been designed by BeagleBoard.org and are manufactured by Circuitco in Richardson, Texas. Both BeagleBoard.org and Circuitco are interested in manufacturing Cape designs submitted by the BeagleBoard.org community.

1.8 POWER REQUIREMENTS AND CAPABILITIES

BeagleBone may be powered from the USB cable, an external 5 VDC power supply or via the Battery Cape. When powered via the USB cable, the original BeagleBone operates at 500 MHz and BeagleBone Black at 1 GHz. When DC powered via an external source, the original BeagleBone operates up to 720 MHz and the Black up to 1 GHz. An external 5 VDC supply may be connected to the external power supply connector. The 5 VDC supply requires a minimum 1 amp current rating and must be equipped with a 2.1 mm center positive with a 5.5 mm outer barrel [Coley]. A BeagleBone–compatible 5 VDC, 2A power supply is available from Adafruit (`www.adafruit.com`).

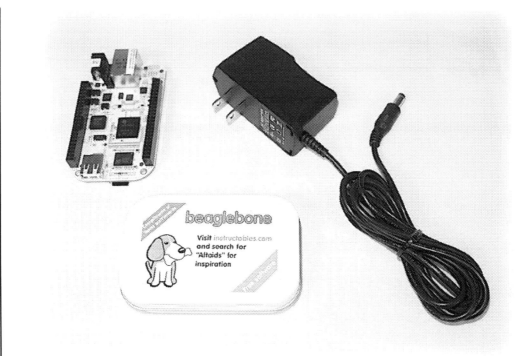

Figure 1.4: BeagleBone power sources (photo courtesy of J. Barrett, Closer to the Sun International).

1.9   GETTING STARTED — SUCCESS OUT OF THE BOX

It is important to the BeagleBoard designers that a novice user experience out–of–box success on first time use. In this section we provide a series of exercises to quickly get up and operating with BeagleBone and Bonescript.

1.9.1   EXERCISE 1: PROGRAMMING WITH BONESCRIPT THROUGH YOUR BROWSER

For the novice user, the quickest way to get up and operating with Bonescript is through your web browser. Detailed step–by–step quick–start instructions are provided in the online "BeagleBone Quick–Start Guide" available at `www.beagleboard.org`. The guide consists of four easy to follow steps. The four steps are summarized below. They may be accomplished in less than 10 minutes to allow rapid out–of–box operation [`www.beagleboard.org`].

1. Plug BeagleBone into the host computer via the mini–USB capable and open the START.htm file, or README.htm for older software images.

2. Install the proper drivers for your system. MS Windows users need to first determine if the host computer is running 32 or 64–bit Windows. Instructions to do this are provided at http://support.microsoft.com/kb/827218.

3. For the original BeagleBone, EJECT the drive original (if you have not updated to the latest Bonescript software image).

4. Browse to "Information on BeagleBoard" information using Chrome or Firefox by going to http://192.168.7.2

5. Explore the Cloud9 IDE develop environment by navigating to http://192.168.7.2:3000/ using Chrome or Firefox.

1.9.2 EXERCISE 2: BLINKING AN LED WITH BONESCRIPT

In this exercise we blink a light emitting diode (LED) onboard BeagleBone and also an external LED. Bonescript will be used along with the blinkled.js program. The LED interface to the BeagleBone is shown in Figure 1.5. The interface circuit consists of a 270 ohm resistor in series with the LED. This limits the current from BeagleBone to a safe value below 8 mA. The LED has two terminals: the anode (+) and the cathode (−). The cathode is the lead closest to the flattened portion of the LED.

The blinkled.js program is provided in the code listing below. The purpose of this program is to blink two different LEDs: on BeagleBone (LED USR3) and an external LED via header P8 pin 13. Only a brief description is provided. A detailed description of Bonescript is provided in the next chapter. In the first line of the code, the Bonescript library must be included. Several lines then follow to allow the program to work in version 0.0.1 (original) and 0.2.0 (black) of Bonescript. The pin designators are then assigned to variable names. The pins are then configured as general purpose outputs. The initial state of the LEDs are set to LOW turning the LEDs off. A JavaScript timer is then configured that runs the function "toggle" once every second (1000 ms). Within the function, the variable "state" is toggled between LOW and HIGH, then used to set the state of the pins driving the LEDs.

```
//*****************************************************
var b = require('bonescript');

//Old bonescript defines 'bone' globally
var pins = (typeof bone != 'undefined') ? bone : b.bone.pins;

var ledPin = pins.P8_13;
var ledPin2 = pins.USR3;

b.pinMode(ledPin, b.OUTPUT);
```

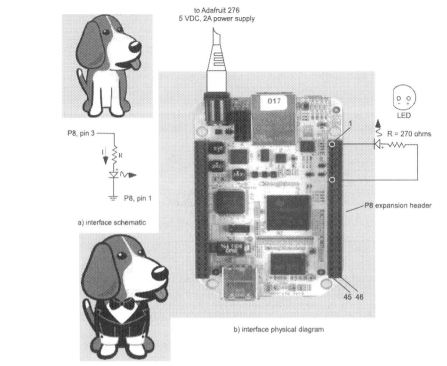

Figure 1.5: Interfacing an LED to BeagleBone. **Notes:** The current being supplied by the BeagleBone pin should not exceed 8 mA.

```
b.pinMode(ledPin2, b.OUTPUT);

var state = b.LOW;
b.digitalWrite(ledPin, state);
b.digitalWrite(ledPin2, state);

setInterval(toggle, 1000);

function toggle() {
   if(state == b.LOW) state = b.HIGH;
   else state = b.LOW;
   b.digitalWrite(ledPin, state);
   b.digitalWrite(ledPin2, state);
}
//*********************************************************
```

1.9.3   EXECUTING THE BLINKLED.JS PROGRAM

To execute the program, Cloud9 IDE is started as described earlier in the chapter. The blinkled.js program is selected and the "run" button is depressed. The LEDs will commence blinking!

1.9.4   EXERCISE 3: DEVELOPING YOUR OWN BONEYARD — AROO!

In this exercise you develop a prototype board to exercise BeagleBone. We have dubbed this the "Boneyard." In the spirit of "Do–it–yourself (DIY)," we encourage you to develop your own Boneyard by using your creativity and sense of adventure.

    If your schedule does not permit time to design and build your own Boneyard, "way cool" prototype boards and cases are available from Adafruit and built–to–spec.com. Adafruit (`www.adafruit.com`) provides a plethora of Beagle related products including an Adafruit proto plate and Bone Box pictured in Figure 1.6. Built–to–spec also offers a variety of BeagleBone products including prototype boards and enclosures. Several are shown in Figure 1.7.

Figure 1.6: Sample of Adafruit BeagleBone products [`www.adafruit.com`].

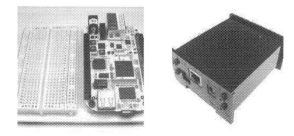

Figure 1.7: Sample of Built–to–spec BeagleBone products `www.built-to-spec.com`].

    For our version of the Boneyard, we have used available off–the–shelf products as shown in Figure 1.8. The Boneyard includes the following:

- A black Pelican Micro Case #1040,

- A BeagleBone evaluation board,

- Two Jameco JE21 3.3 x 2.1 inch solderless breadboards and

- One piece of black plexiglass.

We purposely have not provided any construction details. Instead, we encourage you to use your own imagination to develop and construct your own BeagleBone Boneyard. Also, a variety of Beagle related stickers are available from Café Press www.cafepress.com to customize your boneyard.

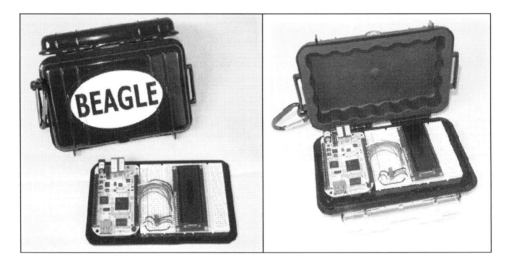

Figure 1.8: BeagleBone Boneyard (photo courtesy of J. Barrett, Closer to the Sun International).

1.10 SUMMARY

This chapter has provided an introduction to BeagleBone, its environment and Beagle community. Also, an overview of BeagleBone features and subsystem was provided. The chapter concluded with several examples to get you started. In the next chapter a detailed introduction to programming and programming tools is provided.

1.11 REFERENCES

- Coley, Gerald. *BeagleBone Rev A6 Systems Reference Manual.* Revision 0.0, May 9, 2012, beagleboard.org www.beagleboard.org

- Dulaney, Emmett. *Linux All—In—One for Dummies.* Hoboken, NJ: Wiley Publishing, Inc., 2010. Print.

- Octavio, "First steps with BeagleBone." `www.borderhack.com`

- "Adafruit Industries." `www.adafruit.com`

- "Built–to–Spec." `www.built-to-spec.com`

1.12   CHAPTER EXERCISES

1. What processor is hosted aboard the original BeagleBone? BeagleBone Black? Summarize the features of each processor.

2. What is the difference between BeagleBone computer and a microcontroller based hobbyist board?

3. What is the operating system aboard BeagleBone? What are the advantages of using this operating system?

4. Describe the Beagle community.

5. Where is the operating system stored on the original BeagleBone? BeagleBone Black?

6. What is the Cloud9 IDE?

7. Summarize the features of Bonescript.

8. Describe the concept of open source hardware.

9. What is a BeagleBone Cape? How many can be used simultaneously? How are conflicts prevented between Capes?

10. Modify the blinkled.js program such that one LED blinks at three times the rate of the other.

CHAPTER 2

System Design: Programming

Objectives: After reading this chapter, the reader should be able to do the following:

- Provide an overview of the system design process and related programming and hardware interfacing challenges.

- Describe the key components of a program.

- Specify the size of different variables within the C programming language.

- Define the purpose of the main program.

- Explain the importance of using functions within a program.

- Write functions that pass parameters and return variables.

- Describe the function of a header file.

- Discuss different programming constructs used for program control and decision processing.

- Outline the key features of programming in JavaScript.

- Describe the key features of the Bonescript library, Node.js interpreter and Cloud9 IDE.

- Describe what features of the Bonescript library, Node.js and Cloud9 IDE ease the program development process.

- Write programs for use on BeagleBone.

- Run BeagleBone "off the leash" – without a host PC connection.

- Apply lessons learned to design and development of an autonomous robot system.

2.1 AN OVERVIEW OF THE DESIGN PROCESS

The first chapter provided an overview of the BeagleBone system, the Bonescript library and the Cloud9 IDE programming environment. The rest of the book is devoted to designing electronic systems with BeagleBone and its accompanying Capes. The next three chapters describe how to design an embedded system using JavaScript and the Bonescript library. Introduction to the JavaScript and C languages is done through comparison and contrast. We apply this information to develop a series of motivating (fun) robots. Following the Bonescript section, the remainder of the book is devoted to electronic systems design using C and open source libraries. We apply this information to the development of a wide variety of devices.

In these next three chapters we design an embedded system using Bonescript by taking a top–down approach. That is, we discuss the pieces of the system first, software programming in Chapter 2 and hardware interfacing in Chapter 3, followed by overall system design approach in Chapter 4. We begin with a brief review of programming basics. Readers who are familiar with JavaScript and C programming may want to skip ahead although a review of the fundamentals may be helpful.

2.2   OVERVIEW

To the novice, programming a processor may appear mysterious, complicated, overwhelming, and difficult. When faced with a new task, one often does not know where to start. The goal of this section is to provide a tutorial on how to begin programming. We will use a top–down design approach. We begin with the "big picture" of the program followed by an overview of the major pieces of a program. We then discuss the basics of the JavaScript and C programming languages. Only the most fundamental programming concepts are covered.

Portions of the programming overview were adapted with permission from earlier Morgan and Claypool projects. Throughout the chapter, we provide examples and also provide pointers to a number of excellent references. The chapter concludes with the development of an autonomous maze navigating robot based on the Blinky 602A platform.

2.3   ANATOMY OF A PROGRAM

Programs have a fairly repeatable format. Slight variations exist but many follow the format provided. We first provide a template for a C program followed by a Node.js program template.

C program template:

```
//****************************************************************
//Comments containing program information
// - file name:
// - author:
// - revision history:
// - compiler setting information:
// - hardware connection description to processor pins
// - program description
//****************************************************************

//include files
#include<file_name.h>
```

```
//function prototypes
A list of functions and their format used within the program

//program constants
#define    TRUE    1
#define    FALSE   0
#define    ON      1
#define    OFF     0

//interrupt handler definitions
Used to link the software to hardware interrupt features

//global variables
Listing of variables used throughout the program

//main program

void main(void)
{

body of the main program

}

//function definitions
A detailed function body and definition for each function
used within the program.
The functions may also include local variables.
//****************************************************************
```

Node.js program template:

```
//****************************************************************
//Comments containing program information
// - file name:
// - author:
// - revision history:
// - interpreter and library requirements:
// - hardware connection description to processor pins
// - program description
```

```
//************************************************************

//include files
var x=require('x');

//function prototypes not required for JavaScript
//because dynamic typing converts variables between types

//no pre-processor means that constants are the same as variables
var TRUE  = 1;
var FALSE = 0;
var ON    = 1;
var OFF   = 0;


//event handler definitions
Used to link the software to asynchronous events

//global variables
Listing of variables used throughout the program

//main program

Unlike C, JavaScript statements don't need to be inside of a
function and will be executed as they are interpreted

body of the main program

//function definitions
A detailed function body and definition for each function
used within the program.
The functions may also include local variables.
//************************************************************
```

For convenience the C and Node.js program templates are shown side–by–side in Figure 2.1. Let's take a closer look at each piece.

2.3.1 COMMENTS

Comments are used throughout the program to document what and how things were accomplished within a program. The comments help you reconstruct your work at a later time. Imagine that you

```
//******************************************     //******************************************
//C program template                            //Node.js program template
//Comments containing program information        Comments containing program information
// - file name:                                  // - file name:
// - author:                                      // - author:
// - revision history:                            // - revision history:
// - compiler setting information:                // - interpreter and library requirements:
// - hardware connection description to           // - hardware connection description to processor pins
//   processor pins                               // - program description
// - program description                         //******************************************
//******************************************
                                                 //include files
//include files                                  var x=require('x');
#include<file_name.h>
                                                 //function prototypes not required for JavaScript
//function prototypes                            //because dynamic typing converts variables between types
A list of functions and their format used within the program
                                                 //no pre-processor means that constants are the same as variables
//program constants                              var TRUE = 1;
#define   TRUE   1                               var FALSE = 0;
#define   FALSE  0                               var ON  = 1;
#define   ON     1                               var OFF  = 0;
#define   OFF    0
                                                 //event handler definitions
//interrupt handler definitions                  Used to link the software to asynchronous events
Used to link the software to hardware interrupt features
                                                 //global variables
//global variables                               Listing of variables used throughout the program
Listing of variables used throughout the program
                                                 //main program
//main program
                                                 Unlike C, JavaScript statements don't need to be inside of a
void main(void)                                  function and will be executed as they are interpreted
{
                                                 body of the main program
body of the main program
                                                 //function definitions
}                                                A detailed function body and definition for each function
                                                 used within the program. The functions may also include local variables.
//function definitions                          //******************************************
A detailed function body and definition for each function
used within the program. The functions may also include local variables.
//******************************************
```

Figure 2.1: C and Node.js program templates.

wrote a program a year ago for a project. You now want to modify that program for a new project. The comments help you remember the key details of the program.

Comments are not compiled into machine code for loading into the processor. Therefore, the comments will not fill up the memory of your processor. Comments are indicated using double slashes (//). Anything from the double slashes to the end of a line is then considered a comment. A multi–line comment can be constructed using a /* at the beginning of the comment and a */ at the end of the comment. These are handy to block out portions of code during troubleshooting or providing multi–line comments. Comment syntax is largely the same between JavaScript and C.

At the beginning of the program, comments may be extensive. Comments may include some of the following information:

- file name

- program author

- revision history or a listing of the key changes made to the program

- instructions on how to compile the program or specific version of interpreter and libraries required

- hardware connection description to processor pins

- program description

2.3.2 INCLUDE FILES

Often you need to add extra files to your project besides the main program. In JavaScript, this is typically where you actually pull in your library code used to perform advanced tasks, such as Bonescript. In C, include files are more typically provide a kind of "personality file" on the specific processor that you are using. This file is provided with the compiler and provides the name of each register used within the processor. It also provides the link between a specific register's name within software and the actual register location within hardware. These files are typically called header files and their name ends with a ".h." You may need to include other header files in your program such as the "math.h" file when programming with advanced math functions.

To include header files within a program, the following syntax is used:

```
//C programming: include files
#include<file_name1.h>
#include<file_name2.h>


//Node.js included libraries
var library1 = require('library1');
var library2 = require('library2');
```

In JavaScript, variables, including functions, declared by a library are typically returned from the call to "require." These variables are typically accessed later using variable member operations.

```
//Node.js included libraries
var library1 = require('library1');

//Node.js access library provided variable memberA using one syntax
var memberA = library1.memberA;

//Node.js access library provided variable memberA using another syntax
var alsoMemberA = library1["memberA"];
```

2.3.3 FUNCTIONS

In Chapter 4, we discuss in detail the top down design, bottom up implementation approach to designing processor based systems. In this approach, a processor based project including both hardware and software is partitioned into systems, subsystems, etc. The idea is to take a complex project and break it into doable pieces with a defined action.

We use the same approach when writing computer programs. At the highest level is the main program which calls functions that have a defined action. When a function is called, program control is released from the main program to the function. Once the function is complete, program control reverts back to the main program.

Functions may in turn call other functions as shown in Figure 2.2. This approach results in a collection of functions that may be reused over and over again in various projects. Most importantly, the program is now subdivided into doable pieces, each with a defined action. This makes writing the program easier but also makes it much easier to modify the program since every action is in a known location.

In C, there are four different pieces of code required to properly configure and call the function:

- the function prototype,

- the function call,

- the function parameter or argument, and

- the function body.

In JavaScript, all of the same pieces of code are required except for the function prototype.

Function prototypes are provided early in the program as previously shown in the program template. The function prototype provides the name of the function and any variables or parameters required by the function and any variable returned by the function.

The function prototype follows this format:

```
return_variable  function_name(required_variable1, required_variable2);
```

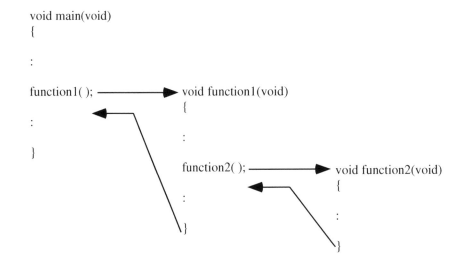

Figure 2.2: Function calling.

If the function does not require variables or does not send back a variable, the word "void" is placed in the variable's position.

The primary purpose of a function prototype is to tell the compiler what types of variables to expect when making or receiving the function call. JavaScript never declares the type of a variable ahead of it being assigned or used. Further, if a JavaScript variable is assigned with one type of variable, say an integer, and then consumed by a function that expects another type of variable, say a string, then JavaScript attempts to automatically convert the variable between the two types. This provides a lot of simplicity to programming as you are learning, but be warned it can get you into trouble as your programs start to get more complex.

The **function call** is the code statement used within a program to execute the function. The function call consists of the function name and the actual arguments required by the function. If the function does not require arguments to be delivered to it for processing, the parenthesis containing the variable list is left empty.

The function call follows this format:

```
function_name(required_variable1, required_variable2);
```

A function that requires no variables follows this format:

```
function_name( );
```

When the function call is executed by the program, program control is transferred to the function, the function is executed, and program control is then returned to the portion of the program that called it.

The syntax for function calls is largely the same between JavaScript and C.

The **function body** is a self–contained "mini–program." In C, the first line of the function body contains the same information as the function prototype: the name of the function, any variables required by the function, and any variable returned by the function. In JavaScript, the return variable type is not provided. Typically, the last line of the function contains a "return" statement. Here a variable may be sent back to the portion of the program that called the function. The processing action of the function is contained within the open ({) and close brackets (}). If the function requires any variables within the confines of the function, they are declared next. These variables are referred to as local variables. A local variable is known only within the scope of a specific function. The local variable is temporarily declared when the function is called and disappears when then function is exited. The actions required by the function follow.

In JavaScript, it is also allowed to have other functions declared as local variables. These functions are special because they utilize the same local "scope" as the other variables declared locally and are not seen directly by outside functions. Because they have access to local variables, it is possible to play some interesting tricks we'll discuss later. It is important to learn about "variable scope" in your development to avoid creating undesired results.

Because JavaScript uses dynamic typing, it is also possible to provide optional arguments to your function and test to see if they are provided within your function body.

The function body in C follows this format:

```
return_type  function_name(required_variable1, required_variable2)
{
//local variables required by the function
unsigned int  variable1;
unsigned char variable2;


//program statements required by the function



//return variable
return return_variable;
}
```

In JavaScript, a function body would look like:

```
function_name(required_variable1, required_variable2, optional_variable3)
{
//local variables required by the function
var variable1;
var variable2;
```

```
//program statements required by the function


//return variable
return return_variable;
}
```

2.3.4 INTERRUPT HANDLER DEFINITIONS

Interrupt service routines are functions that are written by the programmer but usually called by a specific hardware event during system operation. In C, interrupt service routines are handled specially by the compiler because they run outside of the context of the rest of your program. It is not possible to declare actual interrupt service routines in JavaScript, but much of the basic functionality is managed with event handlers. We discuss interrupts and how to properly configure them in Chapter 5.

2.3.5 PROGRAM CONSTANTS

In C, the #define statement is used by the compiler pre-processor to associate a constant name with a numerical value in a program. It can be used to define common constants such as pi. It may also be used to give terms used within a program a numerical value. This makes the code easier to read. For example, the following constants may be defined within a program:

```
//program constants
#define    TRUE    1
#define    FALSE   0
#define    ON      1
#define    OFF     0
```

   JavaScript does not typically make use of any pre-processor and therefore constants must be treated the same as other variables.

2.3.6 VARIABLES

There are two types of variables used within a program: global variables and local variables. A global variable is available and accessible to all portions of the program. Whereas, a local variable is only known and accessible within the context where it is declared.

   When declaring a variable in C, the number of bits used to store the operator is also specified. In Figure 2.3, we provide a list of common C variable sizes. The size of other variables such as pointers, shorts, longs, etc. are contained in the compiler documentation.

   When programming processors, it is important to know the number of bits used to store the variable and correctly assign values to variables. For example, assigning the contents of an unsigned char variable, which is stored in 8–bits, to an 8-bit output port will have a predictable result. However, assigning an unsigned int variable, which is stored in 16–bits, to an 8-bit output

Type	Size	Range
unsigned char	1	0..255
signed char	1	-128..127
unsigned int	2	0..65535
signed int	2	-32768..32767
float	4	+/-1.5e-45.. +/-3.40e+38
double	8	+/-5.0e-324.. +/-1.70e+308

Figure 2.3: Common C variable sizes in bytes.

port does not produce predictable results. It is wise to insure your assignment statements are balanced for accurate and predictable results. The modifier "unsigned" indicates all bits will be used to specify the magnitude of the argument. Signed variables will use the left most bit to indicate the polarity ($\pm$) of the argument.

A variable is declared using the following format. The type of the variable is specified, followed by its name, and an initial value if desired.

```
//global variables
unsigned int  loop_iterations = 6;
```

2.3.7   MAIN FUNCTION

The main function is the hub of activity for the entire program. The main function typically consists of program steps and function calls to initialize the processor followed by program steps to collect data from the environment external to the processor, process the data and make decisions, and provide external control signals back to the environment based on the data collected.

In JavaScript, the main function simply lives in the body of the JavaScript file specified when you invoke the interpreter. For libraries, it is important to avoid having statements outside of variable and function declarations to avoid having those statements invoked when reading in your library.

2.4   FUNDAMENTAL PROGRAMMING CONCEPTS

In the previous section, we covered many fundamental concepts. In this section we discuss operators, programming constructs, and decision processing constructs to complete our fundamental overview of programming concepts.

2.4.1   OPERATORS

There is a wide variety of operators provided in the JavaScript and C languages. An abbreviated list of common operators are provided in Figures 2.4 and 2.5. The operators have been grouped by general category. The symbol, precedence, and brief description of each operator are provided. The precedence column indicates the priority of the operator in a program statement containing multiple operators. Only the fundamental operators are provided. For more information on this topic, see Barrett and Pack in the Reference section at the end of the chapter. JavaScript largely mimics C in terms of operations and precedence.

General operations

Within the general operations category are brackets, parentheses, and the assignment operator. We have seen in an earlier example how bracket pairs are used to indicate the beginning and end of the main program or a function. They are also used to group statements in programming constructs and decision processing constructs. This is discussed in the next several sections.

The parentheses are used to boost the priority of an operator. For example, in the mathematical expression $7 \times 3 + 10$, the multiplication operation is performed before the addition since it has a higher precedence. Parenthesis may be used to boost the precedence of the addition operation. If we contain the addition operation within parentheses $7 \times (3 + 10)$, the addition will be performed before the multiplication operation and yield a different result than the earlier expression.

The assignment operator $(=)$ is used to assign the argument(s) on the right–hand side of an equation to the left–hand side variable. It is important to insure that the left and the right–hand side of the equation have the same type of arguments. If not, unpredictable results may occur.

Arithmetic operations

The arithmetic operations provide for basic math operations using the variables described in the previous section. As described in the previous section, the assignment operator $(=)$ is used to assign the expression on the right–hand side of an equation to the left–hand side variable.

Example: In this example, a function returns the sum of two unsigned int variables which are passed as arguments to the function "sum_two."

```
unsigned int  sum_two(unsigned int variable1, unsigned int variable2)
{
unsigned int  sum;
```

General		
Symbol	Precedence	Description
{ }	1	Brackets, used to group program statements
( )	1	Parenthesis, used to establish precedence
=	12	Assignment

Arithmetic Operations		
Symbol	Precedence	Description
*	3	Multiplication
/	3	Division
+	4	Addition
-	4	Subtraction

Logical Operations		
Symbol	Precedence	Description
<	6	Less than
<=	6	Less than or equal to
>	6	Greater
>=	6	Greater than or equal to
==	7	Equal to
!=	7	Not equal to
&&	9	Logical AND
||	10	Logical OR

Figure 2.4: C operators. (Adapted from [Barrett and Pack]).

Bit Manipulation Operations		
Symbol	Precedence	Description
<<	5	Shift left
>>	5	Shift right
&	8	Bitwise AND
^	8	Bitwise exclusive OR
|	8	Bitwise OR

Unary Operations		
Symbol	Precedence	Description
-	2	Unary negative
~	2	One's complement (bit-by-bit inversion)
++	2	Increment
--	2	Decrement
type(argument)	2	Casting operator (data type conversion)

Figure 2.5: C operators (continued). (Adapted from [Barrett and Pack]).

```
sum = variable1 + variable2;

return sum;
}
```

Logical operations
The logical operators provide Boolean logic operations and are useful in comparing two variables. One argument is compared against another using the logical operator provided. The result is returned as a logic value of one (1, true, high) or zero (0 false, low). The logical operators are used extensively in program constructs and decision processing operations to be discussed later.

Bit manipulation operations
There are two general types of operations in the bit manipulation category: shifting operations and bitwise operations. Let's examine several examples:

Example: Given the following code segment, what will be the value of variable2 after execution?

```
unsigned char    variable1 = 0x73;
unsigned char    variable2;
```

```
variable2 = variable1 << 2;
```

Answer: Variable "variable1" is declared as an eight bit unsigned char and assigned the hexadecimal value of $(73)_{16}$. In binary this is $(0111\_0011)_2$. The $<<\ 2$ operator provides a left shift of the argument by two places. After two left shifts of $(73)_{16}$, the result is $(cc)_{16}$ and will be assigned to the variable "variable2."

Note that the left and right shift operations are equivalent to multiplying and dividing the variable by a power of two.

The bitwise operators perform the desired operation on a bit–by–bit basis. That is, the least significant bit of the first argument is bit–wise operated with the least significant bit of the second argument and so on.

Example: Given the following code segment, what will be the value of variable3 after execution?

```
unsigned char    variable1 = 0x73;
unsigned char    variable2 = 0xfa;
unsigned char    variable3;
```

```
variable3 = variable1 & variable2;
```

Answer: Variable "variable1" is declared as an eight bit unsigned char and assigned the hexadecimal value of $(73)_{16}$. In binary, this is $(0111\_0011)_2$. Variable "variable2" is declared as an eight bit unsigned char and assigned the hexadecimal value of $(fa)_{16}$. In binary, this is $(1111\_1010)_2$. The bitwise AND operator is specified. After execution variable "variable3," declared as an eight bit unsigned char, contains the hexadecimal value of $(72)_{16}$.

Unary operations

The unary operators, as their name implies, require only a single argument.

For example, in the following code segment, the value of the variable "i" is incremented. This is a shorthand method of executing the operation "$i\ =\ i\ +\ 1;$"

```
unsigned int    i = 0;
```

```
i++;
```

Example: It is not uncommon in embedded system design projects to have every pin on a processor employed. Furthermore, it is not uncommon to have multiple inputs and outputs assigned to the same port but on different port input/output pins. Some compilers support specific pin reference. Another technique that is not compiler specific is **bit twiddling**. Figure 2.6 provides bit twiddling examples on how individual bits may be manipulated without affecting other bits using bitwise and unary operators [ImageCraft]. Examples are provided for an eight bit register (Reg_A). This concept may be easily extended to registers of other sizes (e.g., 32 bit registers).

Syntax	Description	Example
a | b	bitwise or	Reg_A |= 0x80; // turn on bit 7 (msb)
a & b	bitwise and	if ((Reg_A & 0x81) == 0) // check if both bit 7 and bit 0 are zero
a ^ b	bitwise exclusive or	Reg_A ^= 0x80; // flip bit 7
~a	bitwise complement	Reg_A &= ~0x80; // turn off bit 7

Figure 2.6: Bit twiddling [ImageCraft].

2.4.2   PROGRAMMING CONSTRUCTS

In this section, we discuss several methods of looping through a set of statements. We examine the "for" and the "while" looping constructs.

The **for** loop provides a mechanism for looping through the same portion of code a fixed number of times or while a certain condition is present. The for loop consists of three main parts:

- initialize variables such as the loop counter,

- loop termination testing, and

- update the loop counter, e.g., increment the loop counter.

In the following code fragment, the for loop is executed 10 times.

```
//In C
unsigned int  loop_ctr;

for(loop_ctr = 0; loop_ctr < 10; loop_ctr++)
  {
                  //loop body

  }
//In Javascript
var loop_ctr;
```

```
for(loop_ctr = 0; loop_ctr < 10; loop_ctr++)
  {
                        //loop body

  }
```

The for loop begins with the variable "loop_ctr" equal to 0. During the first pass through the loop, the variable retains this value. During the next pass, the variable "loop_ctr" is incremented by one. This action continues until the "loop_ctr" variable reaches the value of 10. Since the argument to continue the loop is no longer true, program execution continues beyond the closing braces of the for loop.

In the previous example, the for loop counter was incremented at the beginning of each loop pass. The "loop_ctr" variable can be updated by any amount. For example, in the following code fragment the "loop_ctr" variable is increased by three for every pass of the loop.

```
//In C
unsigned int  loop_ctr;

for(loop_ctr = 0; loop_ctr < 10; loop_ctr=loop_ctr+3)
  {
                        //loop body

  }

//In Javascript
var loop_ctr;

for(loop_ctr = 0; loop_ctr < 10; loop_ctr=loop_ctr+3)
  {
                        //loop body
  }
```

The "loop_ctr" variable may also be initialized at a high value and then decremented at the beginning of each pass of the loop.

```
//In C
unsigned int  loop_ctr;

for(loop_ctr = 10; loop_ctr > 0; loop_ctr--)
  {
                        //loop body
```

```
  }
//In Javascript
var loop_ctr;

for(loop_ctr = 10; loop_ctr > 0; loop_ctr--)
  {
                        //loop body


  }
```

As before, the "loop_ctr" variable may be decreased by any numerical value as appropriate for the application at hand.

In JavaScript, it is also possible to loop over the indexes of an array or keys of an object as shown in the example below.

```
//In Javascript
var index;
var myarray = [1, 2, 3];

for(index in myarray)
  {
     myarray[index]++;    //loop body
  }
```

The value in "index" can be used to index into the array or object. Note that it doesn't contain the indexed value itself, but the value used to perform the index.

The **while** loop is another programming construct that allows multiple passes through a portion of code. The while loop will continue to execute the statements within specified as the second argument of the "for" construct remains logically true. The code snapshot below will implement a ten iteration loop. Note how the "loop_ctr" variable is initialized outside of the loop and incremented within the body of the loop. As before, the variable may be initialized to a greater value and then decremented within the loop body.

```
//In C
unsigned int  loop_ctr;

loop_ctr = 0;
while(loop_ctr < 10)
  {
                        //loop body
```

```
    loop_ctr++;
    }

//In Javascript
var loop_ctr;

loop_ctr = 0;
while(loop_ctr < 10)
    {
                            //loop body
    loop_ctr++;
    }
```

Frequently, within a processor application, the program begins with system initialization actions. Once initialization activities are complete, the processor enters a continuous loop. This may be accomplished using the following code fragment.

```
while(1)
    {


    }
```

In JavaScript, this is sort of synchronous programming is typically avoided and often replaced with use of the JavaScript timers. For example, it is possible to use a JavaScript timer to call a function once every second.

```
var mytimer = setInterval(myfunction, 1000);
```

Here "myfunction" is passed to setInterval as the first argument and the second argument is the number of milliseconds between when the timer goes off to call the function. "myfunction" will be called without any arguments provided to it. "mytimer" contains a reference to the timer that can be used to disable the timer using clearInterval. We'll discuss JavaScript timers in a bit more detail later in this chapter.

2.4.3   DECISION PROCESSING

There are a variety of constructs that allow decision making. These include the following:

- the **if** statement,

- the **if–else** construct,

- the **if–else if–else** construct, and the

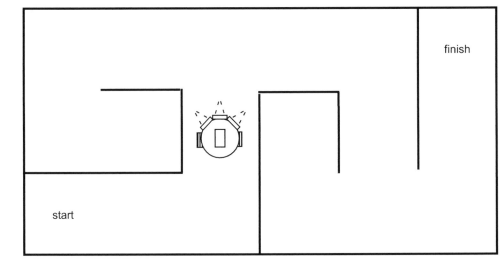

Figure 2.7: Autonomous robot within maze.

- **switch** statement.

The **if** statement will execute the code between an open and close bracket set should the condition within the if statement be logically true.

Example: We use autonomous, maze navigating robots several times throughout the book as electronic systems examples. An autonomous, maze navigating robot is equipped with sensors to detect the presence of maze walls and navigate about the maze. The robot has no prior knowledge about the maze configuration. It uses the sensors and an onboard algorithm to determine the robot's next move. The overall goal is to navigate from the starting point of the maze to the end point as quickly as possible without bumping into maze walls as in Figure 2.7. Maze walls are usually painted white to provide a good, light reflective surface; whereas, the maze floor is painted matte black to minimize light reflections.

In several examples we use the Graymark Blinky 602A robot. This robot platform is quite popular. There are several videos of this robot available online. We refer to it as "Blinky." Suppose our goal is to have Blinky find a path through a maze as shown in Figure 2.7. Blinky is equipped with two wheels driven by DC motors. When moving through the maze, Blinky must safeguard itself from bumping against the walls. For this purpose, we equip Blinky with three IR sensors. The IR sensors provide a voltage output which depends on the distance of the sensor from the reflecting surface.

Later in the chapter we develop the algorithm to allow Blinky to navigate the maze. To help develop the algorithm, a light emitting diode (LED) is connected to a digital input/output pin on BeagleBone. The robot's center infrared (IR) sensor is connected to an analog–to–digital converter (ADC) input pin on BeagleBone. The IR sensor provides a voltage output that is inversely proportional to distance of the sensor from the maze wall. It is desired to illuminate the LED if the robot is within 10 cm of the maze wall. The sensor's output is too large to be directly applied to BeagleBone. We place a voltage divided network between the sensor and BeagleBone to remedy this situation. With the voltage divider network, the sensor provides an output voltage of 1.25 VDC at the 10 cm range. The Bonescript library's analogRead function provides a normalized analog voltage reading from 0 to 1. A reading of 1 corresponds to the maximum ADC system voltage of 1.8 VDC. Therefore, the ADC will report a reading of 0.694 (1.25 VDC/1.80 VDC) when the maze wall is 10 cm from the robot.

The following **if** statement construct will implement this LED indicator. We use pseudocode to illustrate the concept. We provide the actual code to do this later in the chapter.

```
if (center_sensor_output > 0.694) //Center IR sensor voltage
                                  //greater than 1.25 VDC
    {
    led_pin = logic_high;         //illuminate LED connected
                                  //to led_pin
    }
```

In the example provided, there is no method to turn off the LED once it is turned on. This will require the **else** portion of the construct as shown in the next code fragment.

```
if (center_sensor_output > 0.694) //Center IR sensor voltage
                                  //greater than 1.25 VDC
    {
    led_pin = logic_high;         //illuminate LED connected
                                  //to led_pin
    }
else
    {
    led_pin = logic_low;          //turn LED off
    }
```

The **if–else if–else** construct may be used to implement a three LED system. In this example, the left, center, and right IR sensors are connected to three different analog–to–digital converter pins on BeagleBone. Also, three different LEDs are connected to digital output pins on BeagleBone. The following pseudocode fragment implements this LED system.

```
if (left_sensor_output > 0.694)        //Left IR sensor voltage
                                       //greater than 1.25 VDC
```

```
    {
    left_led_pin = logic_high;              //illuminate LED connected
                                            //to left_led_pin

    }

else if (center_sensor_output > 0.694) //Center IR sensor voltage
                                        //greater than 1.25 VDC
    {
    center_led_pin = logic_high;            //illuminate LED connected
                                            //to center_led_pin

    }

else if (right_sensor_output > 0.694)  //Right IR sensor voltage
                                       //greater than 1.25 VDC
    {
    right_led_pin = logic_high;             //illuminate LED connected
                                            //to right_led_pin

    }



else
    {
    left_led_pin = logic_low;               //turn LEDs off
    center_led_pin = logic_low;
    right_led_pin = logic_low;
    }
```

The **switch** statement is used when multiple if–else conditions exist. Each possible condition is specified by a case statement. When a match is found between the switch variable and a specific case entry, the statements associated with the case are executed until a **break** statement is encountered. In general, the break terminates the execution of the nearest enclosing do, switch or while statement and program control passes to the next program statement following the break. In the switch statement, this ensures only a single case of the switch statement is executed.

Example: Suppose an eight bit variable "robot_status" is periodically updated to reflect the current status of the robot (e.g., low battery power, maze walls in the robot's path, etc.). Each bit in the register represents a different robot status item. In response to a change the status the robot must complete status related items. A switch statement may be used to process the multiple possible actions in an orderly manner.

```
if(new_robot_status != old_robot_status)    //check for status change

switch(new_robot_status)
    {                                       //process change in status
    case 0x01:                              //new_robot_status bit 0
                                            //related actions

            break;

    case 0x02:                              //new_robot_status bit 1
                                            //related actions

            break;

    case 0x04:                              //new_robot_status bit 2
                                            //related actions

            break;

    case 0x08:                              //new_robot_status bit 3
                                            //related actions

            break;

    case 0x10:                              //new_robot_status bit 4
                                            //related actions

            break;

    case 0x20:                              //new_robot_status bit 5
                                            //related actions

            break;

    case 0x40:                              //new_robot_status bit 6
                                            //related actions

            break;

    case 0x80:                              //new_robot_status bit 7
                                            //related actions

            break;

            default:;                       //all other cases
    }                                       //end switch(new_robot_status)
```

```
    }                                      //end if new_robot_status
old_robot_status = new_robot_status;       //update old_robot_status
```

That completes our brief overview of the JavaScript and C programming languages. In the next section, we provide a bit more detail on the Bonescript library and the environment in which we will be using it. You will see how this development environment provides a user–friendly method of quickly developing code applications for BeagleBone. But first, let's discuss some of the basics of programming our chosen JavaScript interpreter.

2.5   PROGRAMMING IN JAVASCRIPT USING NODE.JS

In this section we provide an admittedly brief introduction to Node.js. However, we provide pointers to several excellent sources on these topics at the end of the chapter. Node.js is an important topics since Bonescript is written specifically for it. As you will recall, Bonescript is the user–friendly "website" interface to write application programs for BeagleBone. Since JavaScript was specifically developed to quickly implement websites, it was a natural choice to implement Bonescript. JavaScript is an interpreted language. This means the code is not compiled. Instead a JavaScript program is a script or listing of pre–built functions to quickly implement dynamic user interactions within a website.

Node.js is an implementation of a JavaScript interpreter for running on the web host, rather than within your web browser. It was developed to implement event driven programming techniques. In our discussion of Node.js we use a restaurant example to gain a general understanding of event driven programming. Having a fundamental understanding of JavaScript and Node.js will allow you to extend the features and capabilities of Bonescript. It is important to emphasize that Bonescript is an open source library. We are counting on the user community to expand the features of Bonescript. If there is a feature you need, please develop it and share it with the BeagleBoard.org community.

2.5.1   JAVASCRIPT

As previously illustrated, JavaScript is very similar in syntax to the C programming language. It consists of two basic parts: the syntax and the object model. The syntax can be divided into six basic areas:

- comments

- conditionals

- loops

- operators

- functions

- variables

The first four items are virtually identical to the C programming language previously discussed. We discussed function writing and the declaration of variables in the section on Bonescript. It should be noted that strong typed variables are available in JavaScript 2.0 [Vander Veer, Pollock, Kiessling, Hughes–Crocher].

The object model for JavaScript within web browsers is the document object module (DOM). The components of a webpage are referred to as self–contained or encapsulated modules. Module encapsulation includes the object's data, properties, methods and event handlers. Predefined modules are available in HTML, web browsers and JavaScript. Also, custom modules may be implemented using JavaScript[Vander Veer].

2.5.2   EVENT–DRIVEN PROGRAMMING

Node.js uses the same syntax as other JavsScript interpreters, but provides a different object model for running outside of a browser environment. It was developed to execute event–driven programming. Before discussing event driven programming, let's take an aside and discuss breakfast burritos!

Aside. The first author of the book (sfb) is an aficionado of breakfast burritos! The best I've had was from a small family owned restaurant in Colorado Springs, Colorado, called Rita's. I left this beloved city when I retired from military service and moved to Laramie, Wyoming. Laramie is a wonderful place to live and has a wonderful restaurant named Almanza's that serves awesome breakfast burritos. I usually go on Saturdays for this breakfast treat. Provided in Figure 2.8 is a floor plan of the Almanza's restaurant.

On Saturday mornings there is a single server working at Alamanza's. The server is designated at "S1" in the diagram. The server is very efficient and is able to keep up with the steady stream of customers $(C1, C2, \dots)$ that come to the front counter to place their orders and those at the drive–up window $(A1, A2, \dots)$. I really admire those that serve in the restaurant industry. I attempted this earlier in my life and was horrible at it.

Customers would be lost if the server would take the order from counter customer C1 and not serve any other customers until C1's order was complete and delivered. Instead, the server quickly and efficiently takes customer C1's order and passes the order back to the kitchen. Customer C1 is then asked to step aside and wait for their order to be completed. In the meantime the server (S1) takes orders from counter customer C2 and drive up window customer A1. The kitchen then notifies the server S1 that customer C1 order is ready. The server then delivers the food to customer C1.

2.5.3   NODE.JS

The Node.js object model is specifically focused on event driven programming. In contrast, procedural programming is illustrated in Figure 2.9a). In procedural programming a program executes in an orderly fashion as prescribed by the program steps. In event driven programming, the program processes through an event loop as shown in Figure 2.9b). When an asynchronous event occurs,

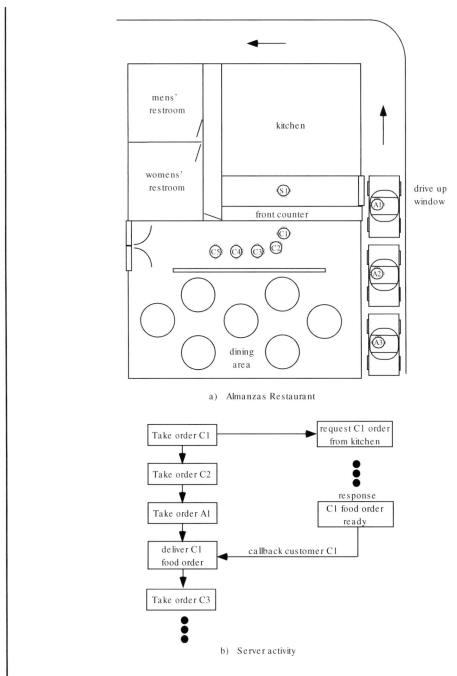

a) Almanzas Restaurant

b) Server activity

Figure 2.8: Alamanza's Restaurant.

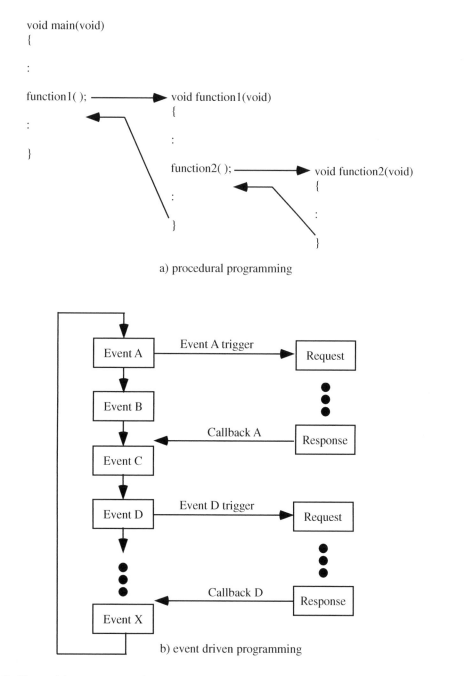

Figure 2.9: Event driven programming.

a **function** is sent to service the event request. In response to the request, event related tasks are performed. If some time is required to complete the event related tasks, the system does not wait for the task to be completed. Instead, the tasks are initiated. When the tasks are complete, the function sent to the activity, the response function termed the callback, is executed. This allows for the efficient processing of multiple activities. The literature describes this as "everything runs in parallel, except your code." That is, the processor is sequentially executing operations. However, event level techniques allow efficient execution of multiple events [Vander Veer, Pollock, Kiessling, Hughes--Crocher]. If this sounds familiar, this is because it is exactly what the server does at Almanza's restaurant. Provided below is an example to better illustrate event driven programming using the Almanza's scenario.

```
//******************************************************************
// almanzas.js
//******************************************************************

var counter = 1;
var driveup = 1;

setTimeout(counterOrder, 1000);//Counter customer at second 1
setTimeout(counterOrder, 2000);//Counter customer at second 2
setTimeout(driveupOrder, 3000);//Driveup customer at second 3
setTimeout(counterOrder, 4000);//Counter customer at second 4
setTimeout(counterOrder, 5000);//Counter customer at second 5
setTimeout(driveupOrder, 6000);//Driveup customer at second 6
setTimeout(driveupOrder, 7000);//Driveup customer at second 7
setTimeout(counterOrder, 8000);//Counter customer at second 8


//******************************************************************
//server(task)
//
//The server takes the orders from the customers and gives them
//to the kitchen, as well as delivering orders from the kitchen
//******************************************************************

function server(customer)
{
    console.log("Take order " + customer);
    kitchen(deliver);

    function deliver()
```

```
    {
        console.log("Deliver " + customer + " food order");
    }
}


//*******************************************************************
//kitchen(callback)
//
//The kitchen takes tasks and completes them at a random time
//between 1 and 5 seconds
//*******************************************************************

function kitchen(callback)
{
    var delay = 1000 + Math.round(Math.random() * 4000);
    setTimeout(callback, delay);
}


//*******************************************************************
//counterOrder()
//
//Give order to server
//*******************************************************************

function counterOrder()
{
    server("C" + counter);
    counter++;
}


//*******************************************************************
//driveupOrder()
//
//Give order to server
//*******************************************************************

function driveupOrder()
{
    server("A" + driveup);
```

```
     driveup++;
}
//******************************************************************
```

Now that you've seen the way JavaScript handles asynchronous tasks, what if you want to run a series of tasks sequentially that have long delays between them without blocking all of your event handlers? Unlike many other programming environments, Node.JS highly discourages using operating system threads that would allow other tasks to run while one thread is blocked. In Node.JS, if we decided to have a function pause and watch the clock, no other function can run until that function returns. Performing sequential tasks is made a bit complicated by this, but here's a pattern that you can use. The series of tasks provided are executed sequentially with delays between them and other event handlers are able to run during those delays.

```
//******************************************************************
// next.js - A pattern for calling sequential functions in node.js
//******************************************************************

mysteps(done);

//******************************************************************
//done
//******************************************************************

function done() {
  console.log("done");
}

//******************************************************************
//mysteps
//******************************************************************

function mysteps(callback)
{
//Provide a list of functions to call
var steps =
[
function(){ console.log("i = " + i); setTimeout(next, 15); }, //delay 15 ms
function(){ console.log("i = " + i); setTimeout(next, 5);  }, //delay 5 ms
function(){ callback(); }
];
```

```
//Start at 0
var i = 0;

console.log("i = " + i);
next(); //Call the first function

//Nested helper function to call the next function in 'steps'
function next()
  {
  i++
  steps[i-1]();
  }
}


//*******************************************************************
```

2.6   BONESCRIPT DEVELOPMENT ENVIRONMENT

Bonescript is a user–friendly, easy–to–use library and environment to harness the power and features of BeagleBone. It was designed to allow one to quickly develop electronic systems. Bonescript consists of a library of event driven functions to use different systems aboard BeagleBone and a set of services running at boot-up to expose many of the functions to a web browser using the socket.io Node.js library. A summary of functions is provided in Figure 2.10. A detailed description of each function is provided in Appendix A. We introduce each function on an as–needed basis in upcoming examples.

Bonescript is currently available in two versions:

- Version 0.0.1 shipped with the original BeagleBone

- Version 0.2.0 shipped with BeagleBone Black

The BeagleBone original can be updated to the new version of Bonescript 0.2.0 if desired. [1] Examples provided throughout the book are compatible with both versions of Bonescript.

The Bonescript functions are easily used within the Cloud9 IDE programming environment illustrated in Figure 2.11. As can be seen, the Cloud9 IDE environment provides familiar file handling features.

Programs for BeagleBone are often written within the Cloud9 IDE programming environment using the Bonescript functions. The program is then executed simply by pushing the "Run." Provided below is the basic format of a Bonescript program. We then illustrate program writing with a series of examples.

[1]To update go to `http://beagleboard.org/Getting+Started#update` and scroll down to "Program an SD Card with the latest software."

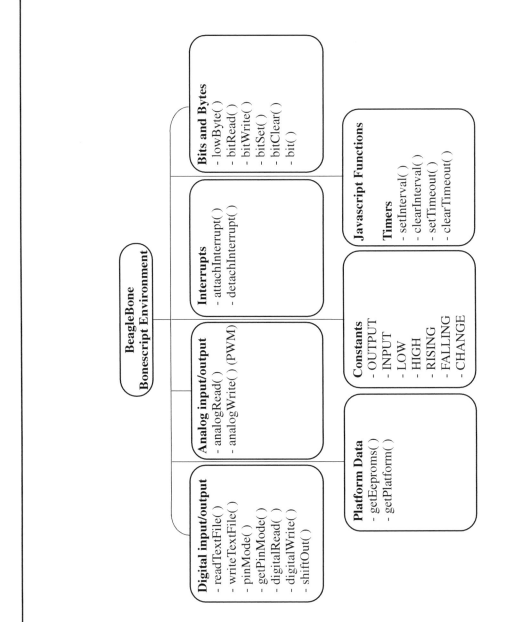

Figure 2.10: Bonescript Development Environment.

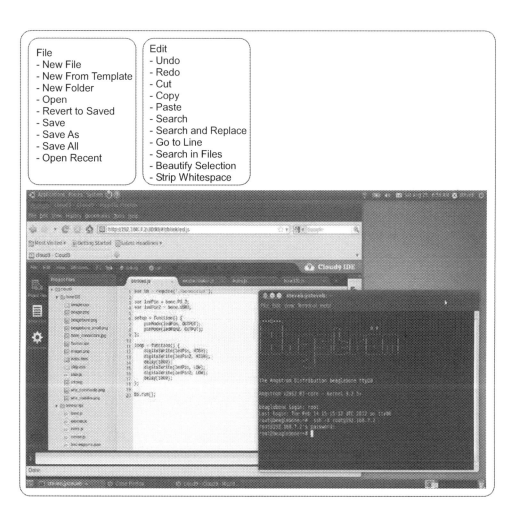

Figure 2.11: Cloud9 IDE programming environment.

```
//********************************************************
var b = require('bonescript');

//define variables used in program
var
   :
   :

//set input and output pin configurations
//initialize required systems
b.pinMode(output_pin, b.OUTPUT);
b.pinMode(input_pin, b.INPUT);

//perform main operation
while(1) {
   :
   :

function1(argument 1, argument 2);

}


//********************************************************
//function1
//********************************************************

function function1(variable 1, variable 2)
{
//function body
   :
   :
}


//********************************************************
```

2.7   APPLICATION 1: ROBOT IR SENSOR

As mentioned earlier in the chapter, we use autonomous, maze navigating robots several times throughout the book as electronic systems examples. An autonomous, maze navigating robot is equipped with sensors to detect the presence of maze walls and navigate about the maze. The robot has no prior knowledge about the maze configuration. It uses the sensors and an onboard algorithm

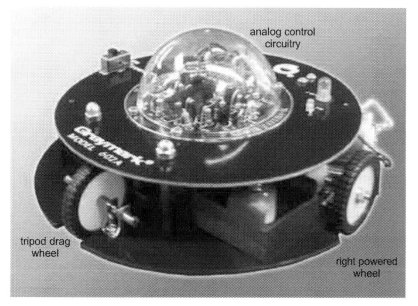

Figure 2.12: Blinky 602A robot. (Figure used with permission of Graymark International Incorporated.)

to determine the robot's next move. The overall goal is to navigate from the starting point of the maze to the end point as quickly as possible without bumping into maze walls as shown in Figure 2.7. Maze walls are usually painted white to provide a good, light reflective surface; whereas, the maze floor is painted matte black to minimize light reflections.

In this first exercise we equip BeagleBone with a single Sharp GP12D IR sensor. This is the same sensor used in the robot. The goal is to become acquainted with the sensor profile and use the sensor to assert a digital output. In the second exercise we use an IR sensor to control a pulse width modulated (PWM) output. This is a good stepping stone to the third exercise. In the third application exercise, we convert a Graymark "Blinky" 602A Pathfinder Robot into a maze navigating robot. The Blinky 602A kit contains the hardware and mechanical parts to construct a line following robot. That is, the Blinky is equipped with downward looking sensors to detect the presence of a black tape line on the floor. It has an analog control system to detect the tape and keep the robot on the tape line. Reference Figure 2.12. The Blinky 602A kit may be purchased from Graymark International, Incorporated (www.graymarkint.com).

We modify the robot platform by equipping it with three Sharp GP12D IR sensors as shown in Figure 2.13. The sensors are available from SparkFun Electronics (www.sparkfun.com). We mount the sensors on a bracket constructed from thin aluminum. Dimensions for the bracket are provided in the figure. Alternatively, the IR sensors may be mounted to the robot platform using "L" brackets available from a local hardware store. In the third application exercise, we equip the robot with all three IR sensors. In this first Application example, we equip the robot with a single sensor and test its function as a proof of concept.

In the first application exercise we want to link a BeagleBone analog input to a digital output. One of the Sharp IR sensors will be connected to an analog input. The IR sensor provides a voltage output as shown in Figure 2.14. The output voltage from the sensor increases and peaks at a specific range and then falls as the range increases.

A software threshold is adjusted such that a digital output will go to logic high when a maze wall is at a desired range. In this application we assume the robot will not be closer than 5 cm to a maze wall. Therefore, we use the portion of the curve beyond the peak where the sensor output is inversely proportional to range. That is, the sensor output voltage decreases with the range from the sensor to maze wall.

It is important to note the input to BeagleBone's analog–to–digital converter may not exceed 1.8 VDC. We use a voltage division circuit between the IR sensor and BeagleBone to stay beneath 1.8 VDC. The interface between the IR sensor and the BeagleBone computer is provided in Figure 2.15. The voltage divider network can be formed using two fixed resistors or a small potentiometer. In the figure a 1MΩ trimmer potentiometer is used to set the maximum value of the IR sensor at 1.75 VDC. The LED has an improved interface circuit using a 2N2222 transistor. This interface circuit will be discussed in Chapter 3.

The IR sensor's power (red wire) and ground (black wire) connections are connected to the 5VDC and ground pins on BeagleBone computer, respectively. The IR sensor's output connection (yellow wire) is connected to an analog input (P9, pin 45) via the trimmer potentiometer on BeagleBone. The LED circuit shown in the top right corner of the diagram is connected to a digital input/output pin (P8, pin 13) on BeagleBone. We discuss the operation of the interface circuit later in the book.

It is desired to illuminate the LED if the robot is within 10 cm of the maze wall. Let's assume the IR sensor interface circuit provides a voltage of 1.25 VDC when a maze wall is 10 cm from the sensor. The code to illuminate the LED at a 10 cm range is provided below.

```
//*********************************************************************

var b=require('bonescript');

//Old bonescript defines 'bone' globally
var pins = (typeof bone != 'undefined') ? bone : b.bone.pins;

var ledPin = pins.P8_13;                    //digital pin for LED interface
var ainPin = pins.P9_39;                    //analog input pin for IR sensor
var IR_sensor_value;
```

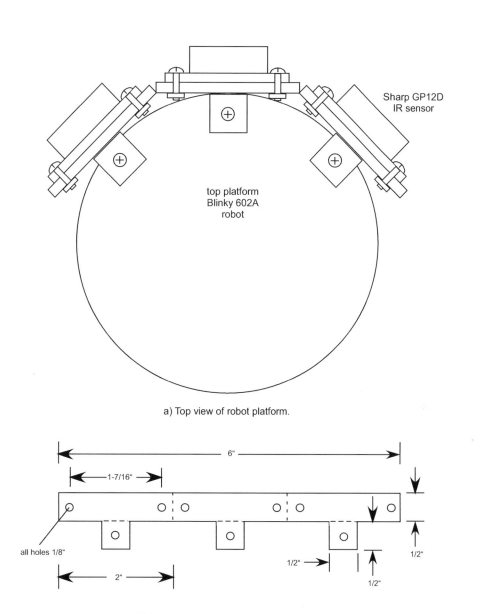

a) Top view of robot platform.

b) Construction details for sensor bracket.

Figure 2.13: Blinky robot platform modified with three IR sensors.

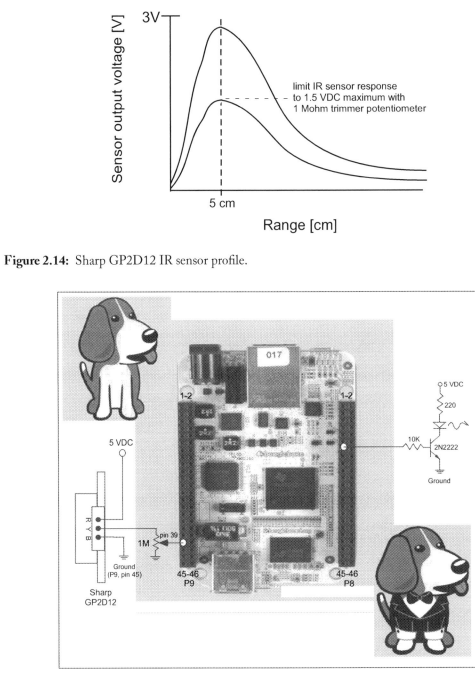

Figure 2.14: Sharp GP2D12 IR sensor profile.

Figure 2.15: IR sensor interface.

```
b.pinMode(ledPin, b.OUTPUT);            //set pin to digital output

while(1)
  {
                                        //read analog output from IR sensor
                                        //normalized value ranges from 0..1
   IR_sensor_value = b.analogRead(ainPin);
                                        //assumes desired threshold at
                                        //1.25 VDC with max value of 1.75 VDC
   if(IR_sensor_value > 0.714)
     {
     b.digitalWrite(ledPin, b.HIGH);   //turn LED on
     }
   else
     {
     b.digitalWrite(ledPin, b.LOW);    //turn LED off
     }
  }
}

//************************************************************************
```

The program begins by providing names for the two BeagleBone board pins that are used in the program. After providing the names for pins, the next step is to declare the pin the LED circuit will use as output within the "setup" function. In this example, the output from the IR sensor will be converted from an analog to a digital value using the built–in Bonescript "analogRead" function. The "analogRead" function requires the pin for analog conversion variable passed to it and returns a normalized value from 0 to 1. The value is normalized to a maximum value of 1.8 VDC.

The loop function calls several functions to read the present analog value on P9, pin 39 and then determine if the reading is above 0.714 (1.25 VDC). If the reading is above 1.25 VDC, the LED on P8, pin 13 is illuminated, else it is turned off.

In the next example, we adapt the IR sensor project to provide custom lighting for an art piece.

2.8   APPLICATION 2: ART PIECE ILLUMINATION SYSTEM

The first author's (sfb) oldest son Jonathan Barrett is a gifted artist (www.closertothesuninternational.com). Although he owns several of his pieces, his favorite one is a painting he did during his early student days. The assignment was to paint your favorite place. Jonathan painted Lac Laronge, Saskatchewan as viewed through the window of a pontoon plane. An image of the painting is provided in Figure 2.16.

Figure 2.16: Lac Laronge, Saskatchewan. Image used with permission, Jonny Barrett, Closer to the Sun Fine Art and Design. (`www.closertothesunfineartndesign.com`)

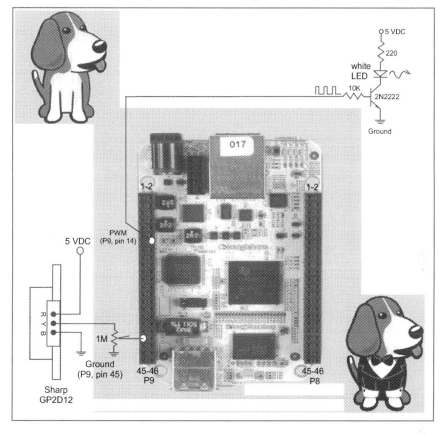

Figure 2.17: IR sensor interface for art illumination.

The circuit provided in the earlier example may be slightly modified to provide custom lighting for an art piece. The IR sensor could be used to detect the presence and position of those viewing the piece. Custom lighting such as a white LED could then be activated by BeagleBone via one of the pulse width modulated (PWM) pins (e.g., P9, pin 14) as shown in Figure 2.17. In the Lac Laronge piece, lighting could be provided from behind the painting using high intensity white LEDs. The intensity of the LEDs could be adjusted by changing the value of PWM duty cycle based on the distance the viewer is from the painting. The PWM concept will be discussed in a later chapter. Briefly, the PWM is a digital signal whose duty cycle (percent on time) may be varied to change the average analog value of the signal. The "analogWrite" function generates a 1 kHz pulse width modulated signal at the specified pin. The duty cycle (the percentage of time the 1 kHz signal is a logic high within a period) is set using a value from 0 to 1. We discuss the PWM concept in greater detail in the next chapter.

```
//**************************************************************************

var b = require('bonescript');

//Old bonescript defines 'bone' globally
var pins = (typeof bone != 'undefined') ? bone : b.bone.pins;

var ledPin = pins.P9_14;              //PWM pin for LED interface
var ainPin = pins.P9_39;              //analog input pin for IR sensor
var IR_sensor_value;

b.pinMode(ledPin, b.OUTPUT);        //set pin to digital output

while(1)
  {
                                    //read analog output from IR sensor
                                    //normalized value ranges from 0..1
  IR_sensor_value = b.analogRead(ainPin);
  b.analogWrite(ledPin, IR_sensor_value);
  }

//**************************************************************************
```

2.9   APPLICATION 3: BLINKY 602A AUTONOMOUS MAZE NAVIGATING ROBOT

In the next several chapters we investigate different autonomous navigating robot designs. Before delving into these designs, it would be helpful to review the fundamentals of robot steering and motor control. Figure 2.18 illustrates the fundamental concepts. Robot steering is dependent upon the number of powered wheels and whether the wheels are equipped with unidirectional or bidirectional control. Additional robot steering configurations are possible. An H–bridge is typically required for bidirectional control of a DC motor. We discuss the H–bridge in greater detail in the next chapter.

2.9.1   BLINKY 602A ROBOT

In this application project we will modify the Blinky 602A robot to be controlled by BeagleBone. The Blinky 602A kit contains the hardware and mechanical parts to construct a line following robot. The processing electronics for the robot consists of analog circuitry. The robot is controlled by two

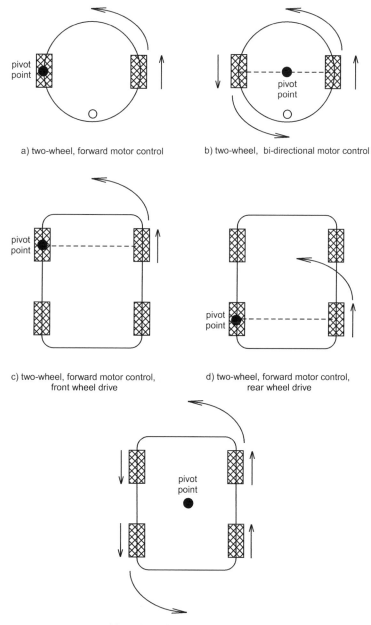

a) two-wheel, forward motor control

b) two-wheel, bi-directional motor control

c) two-wheel, forward motor control,
front wheel drive

d) two-wheel, forward motor control,
rear wheel drive

e) four-wheel, bi-directional motor control

Figure 2.18: Robot control configurations.

3 VDC motors which independently drive a left and right wheel. A third non–powered drag wheel provides tripod stability for the robot.

In this project we equip the Blinky 602A robot platform with three Sharp GP12D IR sensors as shown in Figure 2.19. The characteristics of the sensor were provided earlier in Figure 2.14. The robot is placed in a maze with reflective walls. The project goal is for the robot to detect wall placement and navigate through the maze. It is important to note the robot is not provided any information about the maze. The control algorithm for the robot is hosted on BeagleBone.

2.9.2   REQUIREMENTS

The requirements for this project are simple, the robot must autonomously navigate through the maze without touching maze walls.

2.9.3   CIRCUIT DIAGRAM

The circuit diagram for the robot is provided in Figure 2.20. The three IR sensors (left, center, and right) are mounted on the leading edge of the robot to detect maze walls. The output from the sensors is fed to three separate ADC channels: left – P9, pin 39, center – P9, pin 40, and right – P9, pin 37. The robot motors will be driven by PWM channels: left motor – P9, pin 14 and right motor – P9, pin 16.

BeagleBone is interfaced to the motors via a Darlington transistor (TIP 130, NPN) with enough drive capability to handle the maximum current requirements of the motor.

The robot will be powered by a 9 VDC battery which is fed to a 5 VDC voltage regulator to power BeagleBone. The battery voltage may also be routed to a 3.3 VDC regulator for motor power. Alternatively, a 9 VDC power supply rated at several amps may be used in place of the 9 VDC battery. The supply may be connected to the robot via a flexible umbilical cable.

2.9.4   STRUCTURE CHART

The structure chart for the robot project is provided in Figure 2.21. The structure chart shows the hierarchy of how system hardware and software components will interact and interface with one another. It will be discussed in some detail in Chapter 4.

2.9.5   UML ACTIVITY DIAGRAMS

The UML activity diagram for the robot is provided in Figure 2.22. The activity diagram is simply a UML compliant flow chart. UML is a standardized method of documenting systems. The activity diagram will be discussed in more detail in Chapter 4.

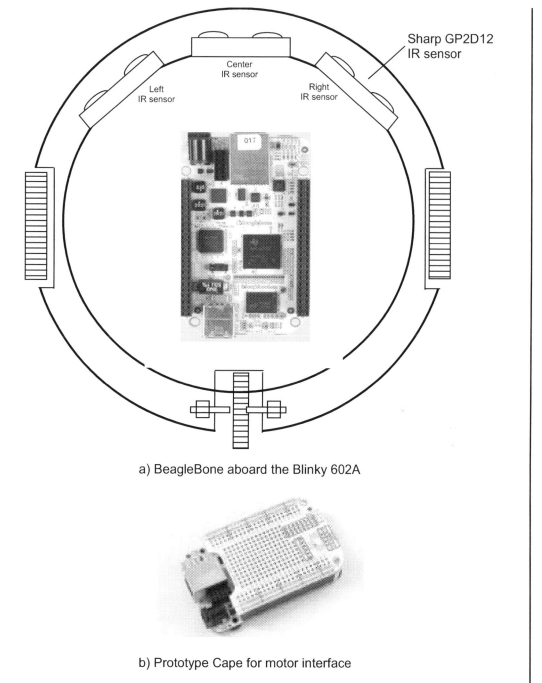

a) BeagleBone aboard the Blinky 602A

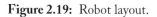

b) Prototype Cape for motor interface

Figure 2.19: Robot layout.

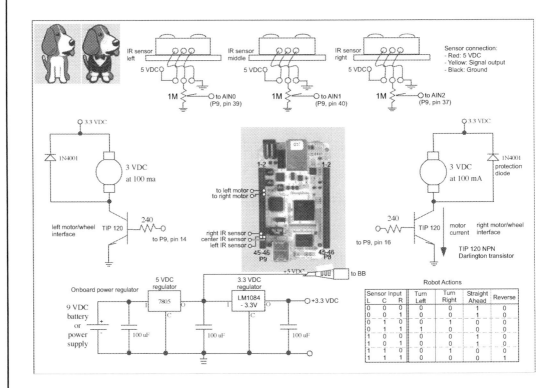

Figure 2.20: Robot circuit diagram.

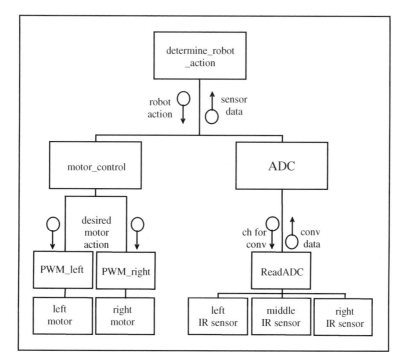

Figure 2.21: Robot structure diagram.

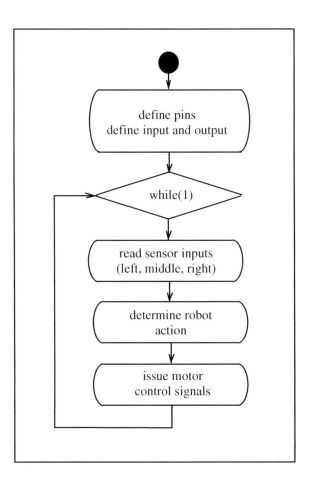

Figure 2.22: Robot UML activity diagram.

2.9.6   BONESCRIPT CODE

Provided below is the basic framework for the Bonescript code. As illustrated in the Robot UML activity diagram, the control algorithm initializes pins, senses wall locations, and issues motor control signals to avoid walls.

It is helpful to characterize the infrared sensor response to the maze walls. This allows a threshold to be determined indicating the presence of a wall. In this example we assume that a threshold of 1.25 VDC has been experimentally determined.

It is important to note that the amount of robot turn is determined by the PWM duty cycle (motor speed) and the length of time the turn is executed. For motors without optical tachometers, the appropriate values for duty cycle and motor on time must be experimentally determined. In the example functions provided the motor PWM and on time are fixed. Functions where the motor duty cycle and on time are passed to the function as local variables are left as a homework assignment.

```
//************************************************************************

var b = require('bonescript');

//Old bonescript defines 'bone' globally
var pins = (typeof bone != 'undefined') ? bone : b.bone.pins;

var left_IR_sensor = pins.P9_39;        //analog input for left IR sensor
var center_IR_sensor = pins.P9_40;      //analog input for center IR sensor
var right_IR_sensor = pins.P9_37;       //analog input for right IR sensor

var left_motor_pin = pins.P9_14;        //PWM pin for left motor
var right_motor_pin = pins.P9_16;       //PWM pin for right motor

var left_sensor_value;
var center_sensor_value;
var right_sensor_value;

b.pinMode(left_motor_pin, b.OUTPUT); //left motor pin
b.pinMode(right_motor_pin, b.OUTPUT);//right motor pin

while(1)
  {
                                //read analog output from IR sensors
                                //normalized value ranges from 0..1
  left_sensor_value   = b.analogRead(left_IR_sensor);
  center_sensor_value = b.analogRead(center_IR_sensor);
```

```
right_sensor_value  = b.analogRead(right_IR_sensor);

                              //assumes desired threshold at
                              //1.25 VDC with max value of 1.75 VDC
if((left_sensor_value > 0.714)&&
   (center_sensor_value <= 0.714)&&
   (right_sensor_value > 0.714))
   {                          //robot continues straight ahead
   b.analogWrite(left_motor_pin, 0.7);
   b.analogWrite(right_motor_pin, 0.7);
   }
else if
   {
   :
   :
   :
   }
}

//*********************************************************************
```

The design provided is very basic. The end of chapter homework assignments extend the design to include the following:

- Modify the PWM turning commands such that the PWM duty cycle and the length of time the motors are on are sent in as arguments to the function.

- Equip the motor with a fourth IR sensor that looks down to the maze floor for "land mines." A land mine consists of a paper strip placed on the maze floor that obstructs a portion of the maze. If a land mine is detected, the robot must "deactivate" it by rotating three times and flashing a large LED while rotating.

- Develop a function for reversing the direction of robot movement.

2.10 SUMMARY

The goal of this chapter was to provide a tutorial on how to begin programming. We used a top-—down design approach. We began with an overview of programming and explained the major parts of JavaScript and C programs for an embedded application. We examined some of the most essential constructs of JavaScript and C programs in greater detail. We then discussed the Bonescript Development Environment, including the Bonescript library, Cloud9 IDE and Node.js JavaScript interpreter, and how it may be used to develop a program for BeagleBone. Throughout the chapter, we provided examples and also provided pointers to a number of excellent references.

2.11 REFERENCES

- "ImageCraft Embedded Systems C Development Tools." `www.imagecraft.com`

- Barrett, Steven and Daniel Pack. *Embedded Systems Design and Applications with the 68HC12 and HCS12.* Upper Saddle River, NJ: Pearson Prentice Hall, 2005. Print.

- Barrett, Jonathan. "Closer to the Sun International." `www.closertothesuninternational.com`

- Barrett, Steven and Daniel Pack. *Processors Fundamentals for Engineers and Scientists.* Morgan and Claypool Publishers, 2006. `www.morganclaypool.com`

- Barrett, Steven and Daniel Pack. *Atmel AVR Processor Primer Programming and Interfacing.* Morgan and Claypool Publishers, 2008. `www.morganclaypool.com`

- Barrett, Steven. *Embedded Systems Design with the Atmel AVR Processor.* Morgan and Claypool Publishers, 2010. `www.morganclaypool.com`

- "ESS Electronic School Supply, Inc." `www.esssales.com`

- "Graymark International, Incorporated." `www.graymarkint.com`

- Vander Veer, Emily. *JavaScript for Dummies.* Hoboken, NJ: Wiley Publishing, Inc., 4th edition, 2005. Print.

- Pollock, John. *JavaScript.* New York, NY: McGraw Hill, 3rd edition, 2010. Print.

- Kiessling, Manuel. *The Node Beginner Guide: A Comprehensive Node.js Tutorial.* 2012. Print.

- Hughes–Croucher, Tom and Mike Wilson. *Node Up and Running.* Sebastopol, CA: O'Reilly Media, Inc., 2012. Print.

2.12 CHAPTER EXERCISES

1. Describe the steps in writing and executing a program on BeagleBone.

2. Describe the key portions of a C program.

3. What is an include file?

4. What are the three pieces of code required for a program function?

5. Describe how to define a program constant.

6. What is the difference between a "for" and "while" loop?

7. When should a switch statement be used versus the if–then statement construct?

8. Describe what variables are require and returned and the basic function of the following built--in Bonescript functions: Blink, Analog Input.

9. Develop a glossary of Linux commands introduced in this chapter.

10. Develop a glossary of Bonescript functions introduced in this chapter.

11. How will you handle the complex profile of the IR sensor?

12. Design a lighting system for an art piece which incorporates three IR sensors and three white LEDs.

13. Complete the Blinky 602A control algorithm for Application 3.

14. Develop a series of functions to turn the Blinky 602A right and left. The motor duty cycle and the motor on time should be passed to the function as a variable.

15. Equip the motor with a fourth IR sensor that looks down to the maze floor for "land mines." A land mine consists of a paper strip placed in the maze floor that obstructs a portion of the maze. If a land mine is detected, the robot must deactivate it by rotating three times and flashing a large LED while rotating.

16. Develop a function for reversing the robot.

CHAPTER 3

BeagleBone Operating Parameters and Interfacing

Objectives: After reading this chapter, the reader should be able to

- Describe the voltage and current parameters for BeagleBone.

- Apply the voltage and current parameters toward properly interfacing input and output devices to BeagleBone.

- Interface BeagleBone operating at 3.3 VDC with a peripheral device operating at 5.0 VDC.

- Interface a wide variety of input and output devices to BeagleBone.

- Describe the special concerns that must be followed when BeagleBone is used to interface to a high power DC or AC device.

- Describe how to control the speed and direction of a DC motor.

- Describe how to control several types of AC loads.

- Summarize the layout of the Prototype Cape for BeagleBone.

- Equip the Blinky 602A robot with a Liquid Crystal Display (LCD).

3.1 OVERVIEW

In this chapter, we introduce the extremely important concepts of the operating envelope for a processor. We begin by reviewing the voltage and current electrical parameters for BeagleBone. We use this information to properly interface input and output devices to BeagleBone. BeagleBone operates at 3.3 VDC. There are many compatible peripheral devices. However, many peripheral devices still operate at 5.0 VDC. We discuss how to interface a 3.3 VDC microcontroller to 5.0 VDC peripherals. We then discuss the special considerations for controlling a high power DC or AC load such as a motor. The overview of interface techniques was adapted with permission from other Morgan and Claypool projects [Pack and Barrett]. Throughout the chapter, we provide a number of detailed examples to illustrate concepts. If this is your first exposure to electronics, please consider reading "Getting Started in Electronics" by Forrest M. Mims III [Mims]. This book provides a well–written, comprehensive introduction to the fascinating world of electronics.

3.2   OPERATING PARAMETERS

A processor is an electronic device which has precisely defined operating conditions. As long as the processor is used within its defined operating parameter limits, it should continue to operate correctly. However, if the allowable conditions are violated, spurious results may arise.

3.2.1   BEAGLEBONE 3.3 VDC OPERATION

Any time a device is connected to a processor, careful interface analysis must be performed. Beagle-Bone digital input/output signals operate at 3.3 VDC. To perform interface analysis, there are eight different electrical specifications we must consider. The electrical parameters are defined below and illustrated in Figure 3.1.

- V_{OH}: the lowest guaranteed output voltage for a logic high,

- V_{OL}: the highest guaranteed output voltage for a logic low,

- I_{OH}: the output current for a V_{OH} logic high,

- I_{OL}: the output current for a V_{OL} logic low,

- V_{IH}: the lowest input voltage guaranteed to be recognized as a logic high,

- V_{IL}: the highest input voltage guaranteed to be recognized as a logic low,

- I_{IH}: the input current for a V_{IH} logic high, and

- I_{IL}: the input current for a V_{IL} logic low.

To properly interface a peripheral device to BeagleBone, the parameters provided in Figure 3.2 must be used. It is important to realize that these are static values taken under very specific operating conditions. If external circuitry is connected such that the processor acts as a current source (current leaving processor) or current sink (current entering processor), the voltage parameters listed above will also be affected.

In the current source case, an output voltage V_{OH} is provided at the output pin of the processor when the load connected to this pin draws a current of I_{OH}. If a load draws more current from the output pin than the I_{OH} specification, the value of V_{OH} is reduced. If the load current becomes too high, the value of V_{OH} falls below the value of V_{IH} for the subsequent logic circuit stage, and it will not be recognized as an acceptable logic high signal. When this situation occurs, erratic and unpredictable circuit behavior results.

In the sink case, an output voltage V_{OL} is provided at the output pin of the processor when the load connected to this pin delivers a current of I_{OL} to this logic pin. If a load delivers more current to the output pin of the processor than the I_{OL} specification, the value of V_{OL} increases. If the load current becomes too high, the value of V_{OL} rises above the value of V_{IL} for the subsequent logic circuit stage, and it will not be recognized as an acceptable logic low signal. As before, when this situation occurs, erratic and unpredictable circuit behavior results.

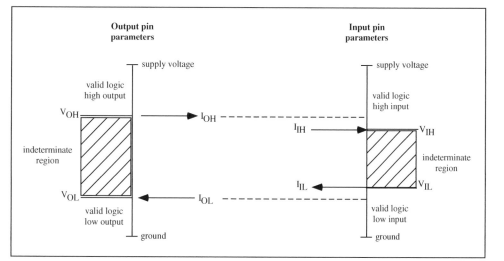

Figure 3.1: Parameters definitions.

3.2.2   COMPATIBLE 3.3 VDC LOGIC FAMILIES

There are several compatible logic families that operate at 3.3 VDC. These families include the LVC, LVA, and the LVT logic families. Key parameters for the low voltage compatible families are provided in Figure 3.3.

3.2.3   INPUT/OUTPUT OPERATION AT 5.0 VDC

BeagleBone operates at 3.3 VDC. However, many HC CMOS microcontroller families and peripherals operate at a supply voltage of 5.0 VDC. For completeness, we provide operating parameters for these types of devices. This information is essential should BeagleBone be interfaced to a 5 VDC CMOS device or peripheral.

Typical values for a microcontroller in the HC CMOS family, assuming V_{DD} = 5.0 volts and V_{SS} = 0 volts, are provided below. The minus sign on several of the currents indicates a current flow out of the device. A positive current indicates current flow into the device.

- V_{OH} = 4.2 volts,

- V_{OL} = 0.4 volts,

- I_{OH} = -0.8 milliamps,

- I_{OL} = 1.6 milliamps,

- V_{IH} = 3.5 volts,

- V_{IL} = 1.0 volt,

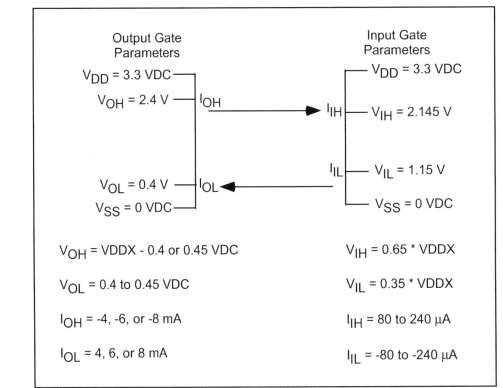

Figure 3.2: BeagleBone interface parameters.

- I_{IH} = 10 microamps, and

- I_{IL} = -10 microamps.

3.2.4   INTERFACING 3.3 VDC LOGIC FAMILIES TO 5.0 VDC LOGIC FAMILIES

Although there are a wide variety of available 3.3 VDC peripheral devices available for BeagleBone, some useful peripheral devices are not available for a 3.3 VDC system. If bidirectional information exchange is required between the processor and the peripheral device, a bidirectional level shifter is required between the processor and the peripheral device. The level shifter translates the 3.3 VDC signal to 5 VDC for the peripheral device and back down to 3.3 VDC for the processor. There are a wide variety of unidirectional and bidirectional level shifting devices available. For example, Sparkfun Electronics has the BOB–08745 logic level shifter breakout board (www.sparkfun.com).

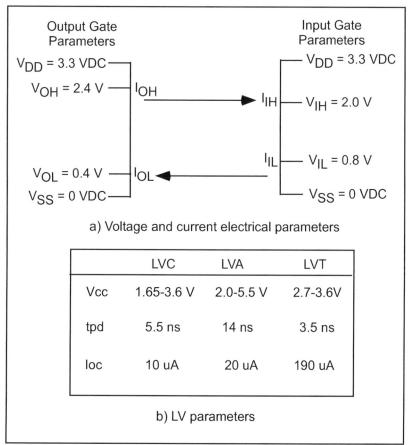

a) Voltage and current electrical parameters

	LVC	LVA	LVT
Vcc	1.65-3.6 V	2.0-5.5 V	2.7-3.6V
tpd	5.5 ns	14 ns	3.5 ns
loc	10 uA	20 uA	190 uA

b) LV parameters

Figure 3.3: Low voltage compatible logic families.

3.3   INPUT DEVICES

In this section, we discuss how to properly interface input devices to a processor. We start with the most basic input component, a simple on/off switch.

3.3.1   SWITCHES

Switches come in many varieties. As a system designer it is up to you to choose the appropriate switch for a specific application. Switch varieties commonly used in processor applications are illustrated in Figure 3.4a). Here is a brief summary of the different types:

- **Slide switch:** A slide switch has two different positions: on and off. The switch is manually moved to one position or the other. For processor applications, slide switches are available that

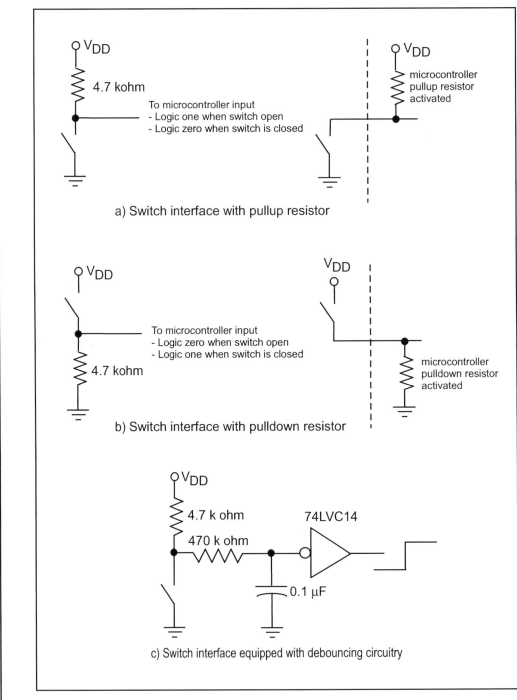

Figure 3.4: Switch interface.

fit in the profile of a common integrated circuit size dual inline package (DIP). A bank of four or eight DIP switches in a single package is commonly available.

- **Momentary contact pushbutton switch:** A momentary contact pushbutton switch comes in two varieties: normally closed (NC) and normally open (NO). A normally open switch, as its name implies, does not normally provide an electrical connection between its contacts. When the pushbutton portion of the switch is depressed, the connection between the two switch contacts is made. The connection is held as long as the switch is depressed. When the switch is released, the connection is opened. The converse is true for a normally closed switch. For processor applications, pushbutton switches are available in a small tactile (tact) type switch configuration.

- **Push on/push off switches:** These type of switches are also available in a normally open or normally closed configuration. For the normally open configuration, the switch is depressed to make connection between the two switch contacts. The pushbutton must be depressed again to release the connection.

- **Hexadecimal rotary switches:** Small profile rotary switches are available for processor applications. These switches commonly have sixteen rotary switch positions. As the switch is rotated to each position, a unique four bit binary code is provided at the switch contacts.

Common switch interfaces are shown in Figure 3.4(b) and (c). This interface allows a logic one or zero to be properly introduced to a processor input port pin. The basic interface consists of the switch in series with a current limiting resistor. The node between the switch and the resistor is provided to the processor input pin. In the configuration shown, the resistor pulls the processor input up to the supply voltage V_{DD}. When the switch is closed, the node is grounded and a logic zero is detected by the processor input pin. To reverse the logic of the switch configuration, the position of the resistor and the switch is simply reversed.

3.3.2   SWITCH DEBOUNCING

Unfortunately mechanical switches do not make a clean transition from one position (on) to another (off). When a switch is moved from one position to another, it makes and breaks contact multiple times. This activity may go on for tens of milliseconds. A processor is relatively fast as compared to the action of the switch. Therefore, the processor is able to recognize each switch bounce as a separate and erroneous transition.

To correct the switch bounce phenomenon, additional external hardware components may be used or software techniques may be employed. A hardware debounce circuit is illustrated in Figure 3.4(c). The node between the switch and the limiting resistor of the basic switch circuit is fed to a low pass filter (LPF), formed by the 470 k Ohm resistor and the capacitor. The LPF prevents abrupt changes (bounces) in the input signal from the processor. The LPF is followed by a 74LVC14

Schmitt Trigger which is simply an inverter equipped with hysteresis. This further limits the switch bouncing.

Switches may also be debounced using software techniques. This is accomplished by inserting a 30 to 50 ms lockout delay in the function responding to port pin changes. The delay prevents the processor from responding to the multiple switch transitions due to bouncing.

You must carefully analyze a given design to determine if hardware or software switch debouncing techniques should be used. It is important to remember that all switches exhibit bounce phenomena and therefore must be debounced.

3.3.3   KEYPADS

A keypad is simply an extension of the simple switch configuration. A typical keypad configuration and interface are shown in Figure 3.5. As you can see, the keypad simply consists of multiple switches in the same package. A hexadecimal keypad is shown in the figure. A single row of keypad switches is asserted by the processor, and then the host keypad port is immediately read. If a switch in this row has been depressed, the keypad pin corresponding to the column the switch is in will also be asserted. The combination of a row and a column assertion can be decoded to determine which key has been pressed as illustrated in the table. Keypad rows are continually asserted one after the other in sequence. Since the keypad is a collection of switches, debounce techniques must also be employed.

The keypad may be used to introduce user requests to a processor. A standard keypad with alphanumeric characters may be used to provide alphanumeric values to the processor such as providing your personal identification number (PIN) for a financial transaction. However, some keypads are equipped with removable switch covers such that any activity can be associated with a key press.

3.3.4   SENSORS

A processor is typically used in control applications where data is collected, the data is assimilated and processed by the host algorithm, and a control decision and accompanying signals are provided by the processor. Input data for the processor is collected by a complement of input sensors. These sensors may be digital or analog in nature.

Digital Sensors

Digital sensors provide a series of digital logic pulses with sensor data encoded. The sensor data may be encoded in any of the parameters associated with the digital pulse train such as duty cycle, frequency, period, or pulse rate. The input portion of the timing system may be configured to measure these parameters.

An example of a digital sensor is an optical encoder. An optical encoder consists of a small transparent, plastic disk with opaque lines etched on the disk surface. A stationary optical emitter and detector source are placed on either side of the disk. As the disk rotates, the opaque lines break the continuity between the optical source and detector. The signal from the optical detector is monitored

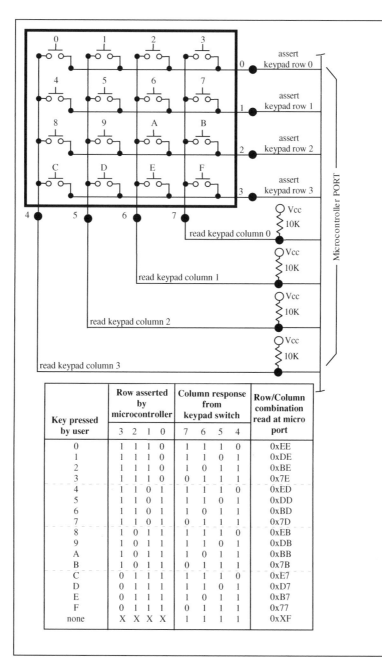

Key pressed by user	Row asserted by microcontroller				Column response from keypad switch				Row/Column combination read at micro port
	3	2	1	0	7	6	5	4	
0	1	1	1	0	1	1	1	0	0xEE
1	1	1	1	0	1	1	0	1	0xDE
2	1	1	1	0	1	0	1	1	0xBE
3	1	1	1	0	0	1	1	1	0x7E
4	1	1	0	1	1	1	1	0	0xED
5	1	1	0	1	1	1	0	1	0xDD
6	1	1	0	1	1	0	1	1	0xBD
7	1	1	0	1	0	1	1	1	0x7D
8	1	0	1	1	1	1	1	0	0xEB
9	1	0	1	1	1	1	0	1	0xDB
A	1	0	1	1	1	0	1	1	0xBB
B	1	0	1	1	0	1	1	1	0x7B
C	0	1	1	1	1	1	1	0	0xE7
D	0	1	1	1	1	1	0	1	0xD7
E	0	1	1	1	1	0	1	1	0xB7
F	0	1	1	1	0	1	1	1	0x77
none	X	X	X	X	1	1	1	1	0xXF

Figure 3.5: Keypad interface.

to determine disk rotation as shown in Figure 3.6. The optical encoder configuration provides an optical tachometer.

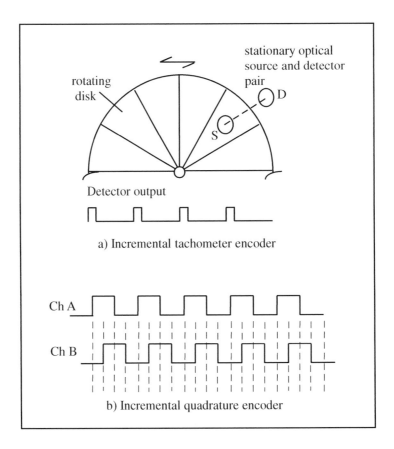

Figure 3.6: Optical encoder.

There are two major types of optical encoders: incremental encoders and absolute encoders. An absolute encoder is used when it is required to retain position information when power is lost. For example, if you were using an optical encoder in a security gate control system, an absolute encoder would be used to monitor the gate position. An incremental encoder is used in applications where a velocity or a velocity and direction information is required.

The incremental encoder types may be further subdivided into tachometers and quadrature encoders. An incremental tachometer encoder consists of a single track of etched opaque lines as shown in Figure 3.6(a). It is used when the velocity of a rotating device is required. To calculate velocity, the number of detector pulses is counted in a fixed amount of time. Since the number of pulses per encoder revolution is known, velocity may be calculated.

The quadrature encoder contains two tracks shifted in relationship to one another by 90 degrees. This allows the calculation of both velocity and direction. To determine direction, one would monitor the phase relationship between Channel A and Channel B as shown in Figure 3.6(b). The absolute encoder is equipped with multiple data tracks to determine the precise location of the encoder disk [Sick Stegmann].

Analog Sensors and Transducers

Analog sensors or transducers provide a DC voltage that is proportional to the physical parameter being measured. The analog signal may be first preprocessed by external analog hardware such that it falls within the voltage references of the conversion subsystem. In the case of BeagleBone, the transducer output must fall between 0 and 1.8 VDC. The analog voltage is then converted to a corresponding binary representation. **Note: As previously mentioned, the analog input to the analog–to–digital converter must not exceed 1.8 VDC.**

Example 1: Flex Sensor

An example of an analog sensor is the flex sensor shown in Figure 3.7(a). The flex sensor provides a change in resistance for a change in sensor flexure. At 0 degrees flex, the sensor provides 10k Ohms of resistance. For 90 degrees flex, the sensor provides 30–40k Ohms of resistance. Since the processor can not measure resistance directly, the change in flex sensor resistance must be converted to a change in a DC voltage. This is accomplished using the voltage divider network shown in Figure 3.7(c). For increased flex, the DC voltage will increase. The voltage can be measured using the BeagleBone's analog–to–digital converter subsystem.

The flex sensor may be used in applications such as virtual reality data gloves, robotic sensors, biometric sensors, and in science and engineering experiments [Images Company]. One of the co-–authors (sfb) used the circuit provided in Figure 3.7 to help a colleague in Zoology monitor the movement of a newt salamander during a scientific experiment.

Example 2: Ultrasonic sensor

The ultrasonic sensor pictured in Figure 3.9 is an example of an analog based sensor. The sensor is based on the concept of ultrasound or sound waves that are at a frequency above the human range of hearing (20 Hz to 20 kHz). The ultrasonic sensor pictured in Figure 3.9c) emits a sound wave at 42 kHz. The sound wave reflects from a solid surface and returns back to the sensor. The amount of time for the sound wave to transit from the surface and back to the sensor may be used to determine the range from the sensor to the wall. Pictured in Figure 3.9(c) and (d) is an ultrasonic sensor manufactured by Maxbotix (LV–EZ3). The sensor provides an output that is linearly related to range in three different formats: a) a serial RS–232 compatible output at 9600 bits per second, b) a pulse width modulated (PWM) output at a 147 us/inch duty cycle, and c) an analog output at a resolution of 10 mV/inch. The sensor is powered from a 2.5 to 5.5 VDC source [www.sparkfun. com].

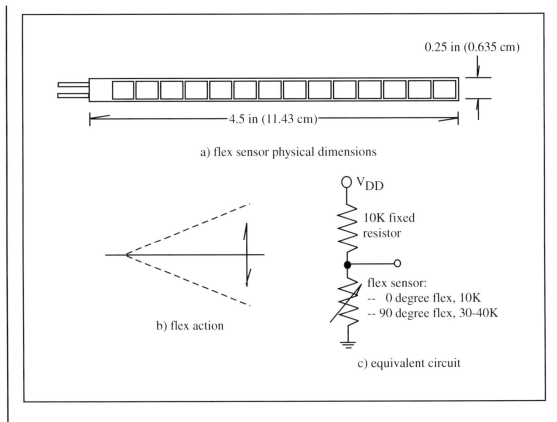

Figure 3.7: Flex sensor.

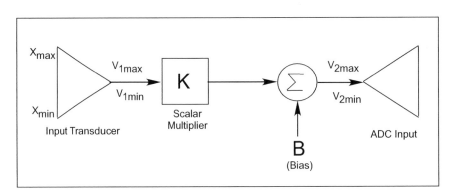

Figure 3.8: A block diagram of the signal conditioning for an analog–to–digital converter. The range of the sensor voltage output is mapped to the analog–to–digital converter input voltage range. The scalar multiplier maps the magnitudes of the two ranges and the bias voltage is used to align two limits.

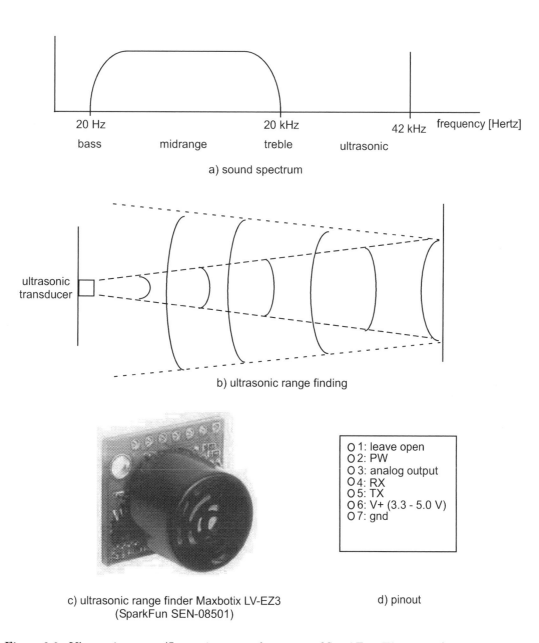

a) sound spectrum

b) ultrasonic range finding

O 1: leave open
O 2: PW
O 3: analog output
O 4: RX
O 5: TX
O 6: V+ (3.3 - 5.0 V)
O 7: gnd

c) ultrasonic range finder Maxbotix LV-EZ3
(SparkFun SEN-08501)

d) pinout

Figure 3.9: Ultrasonic sensor. (Sensor image used courtesy of SparkFun, Electronics.)

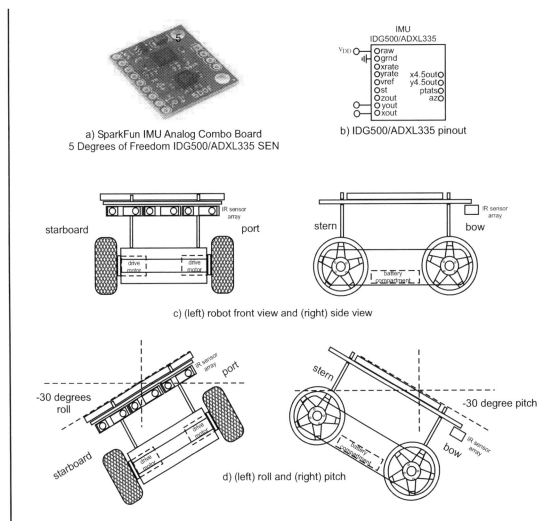

a) SparkFun IMU Analog Combo Board
5 Degrees of Freedom IDG500/ADXL335 SEN

b) IDG500/ADXL335 pinout

c) (left) robot front view and (right) side view

d) (left) roll and (right) pitch

Figure 3.10: Inertial measurement unit. (IMU image used courtesy of SparkFun, Electronics.)

Example 3: Inertial Measurement Unit

Pictured in Figure 3.10 is an inertial measurement unit (IMU) combination which consists of an IDG5000 dual–axis gyroscope and an ADXL335 triple axis accelerometer. This sensor may be used in unmanned aerial vehicles (UAVs), autonomous helicopters and robots. For robotic applications the robot tilt may be measured in the X and Y directions as shown in Figure 3.10(c) and (d) [www.sparkfun.com].

3.3.5   TRANSDUCER INTERFACE DESIGN (TID) CIRCUIT

In addition to transducers, we also need a signal conditioning circuitry before we can apply the signal for analog–to–digital conversion. The signal conditioning circuitry is called the transducer interface. The objective of the transducer interface circuit is to scale and shift the electrical signal range to map the output of the transducer to the input range of the analog–to–digital converter, which is typically 0 to 3.3 VDC. Figure 3.8 shows the transducer interface circuit using an input transducer.

The transducer interface consists of two steps: scaling and shifting via a DC bias. The scale step allows the span of the transducer output to match the span of the analog–to–digital conversion (ADC) system input range. The bias step shifts the output of the scale step to align with the input of the ADC system. In general, the scaling and bias process may be described by two equations:

$$V_{2max} = (V_{1max} \times K) + B$$

$$V_{2min} = (V_{1min} \times K) + B$$

The variable V_{1max} represents the maximum output voltage from the input transducer. This voltage occurs when the maximum value of the physical variable (X_{max}) is presented to the input transducer. This voltage must be scaled by the scalar multiplier (K) and then have a DC offset bias voltage (B) added to provide the voltage V_{2max} to the input of the ADC converter [USAFA].

Similarly, the variable V_{1min} represents the minimum output voltage from the input transducer. This voltage occurs when the minimum physical variable (X_{min}) is presented to the input transducer. This voltage must be scaled by the scalar multiplier (K) and then have a DC offset bias voltage (B) added to produce voltage V_{2min}, the input of the ADC converter.

Usually the values of V_{1max} and V_{1min}, are provided with the documentation for the transducer. Also, the values of V_{2max} and V_{2min} are known. They are the high and low reference voltages for the ADC system (1.8 VDC and 0 VDC for BeagleBone). We thus have two equations and two unknowns to solve for K and B. The circuits to scale by K and add the offset B are usually implemented with operational amplifiers.

Example: A photodiode is a semiconductor device that provides an output current, corresponding to the light impinging on its active surface. The photodiode is used with a transimpedance amplifier to convert the output current to an output voltage. A photodiode/transimpedance amplifier provides an output voltage of 0 volts for maximum rated light intensity and -2.50 VDC output voltage for the minimum rated light intensity. Calculate the required values of K and B for this light transducer, so it may be interfaced to BeagleBone's ADC system.

$$V_{2max} = (V_{1max} \times K) + B$$

$$V_{2min} = (V_{1min} \times K) + B$$

$$1.8 \ V \ = \ (0 \ V \ \times \ K) \ + \ B$$

$$0 \ V \ = \ (-2.50 \ V \ \times \ K) \ + \ B$$

The values of K and B may then be determined to be 0.72 and 1.8 VDC, respectively.

3.3.6   OPERATIONAL AMPLIFIERS

In the previous section, we discussed the transducer interface design (TID) process. Going through this design process yields a required value of gain (K) and DC bias (B). Operational amplifiers (op amps) are typically used to implement a TID interface. In this section, we briefly introduce operational amplifiers including ideal op amp characteristics, classic op amp circuit configurations, and an example to illustrate how to implement a TID with op amps. Op amps are also used in a wide variety of other applications, including analog computing, analog filter design, and a myriad of other applications. Interested readers are referred to the References section at the end of the chapter for pointers to some excellent texts on this topic.

The ideal operational amplifier

An ideal operational amplifier is shown in Figure 3.11. An ideal operational amplifier does not exist in the real world. However, it is a good first approximation for use in developing op amp application circuits.

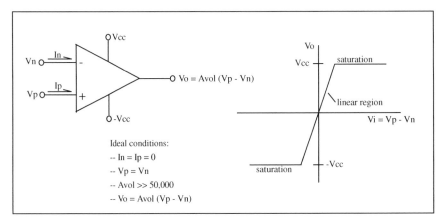

Figure 3.11: Ideal operational amplifier characteristics.

The op amp is an active device (requires power supplies) equipped with two inputs, a single output, and several voltage source inputs. The two inputs are labeled Vp, or the non–inverting input, and Vn, the inverting input. The output of the op amp is determined by taking the difference between Vp and Vn and multiplying the difference by the open loop gain (A_{vol}) of the op amp, which is typically 10,000. Due to the large value of A_{vol}, it does not take much of a difference between Vp and Vn before the op amp will saturate. When an op amp saturates, it does not damage the op amp, but the output is limited to $\pm V_{cc}$. This will clip the output, and hence distort the signal, at levels slightly less than $\pm V_{cc}$. Op amps are typically used in a closed loop, negative feedback configuration. A sample of classic operational amplifier configurations with negative feedback are provided in Figure 3.12 [Faulkenberry].

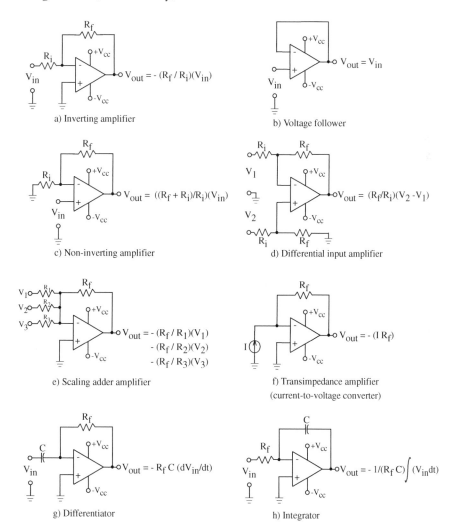

Figure 3.12: Classic operational amplifier configurations. Adapted from [Faulkenberry].

It should be emphasized that the equations provided with each operational amplifier circuit are only valid if the circuit configurations are identical to those shown. Even a slight variation in the circuit configuration may have a dramatic effect on circuit operation. To analyze each operational amplifier circuit, use the following steps:

- Write the node equation at the inverting input.

- Apply ideal op amp characteristics to the node equation.

- Solve the node equation for Vo.

As an example, we provide the analysis of the non–inverting amplifier circuit in Figure 3.13. This same analysis technique may be applied to all of the circuits in Figure 3.12 to arrive at the equations for Vout provided.

Example: In the previous section, it was determined that the values of K and B were 0.72 and 1.8 VDC, respectively. The two–stage op amp circuitry in Figure 3.14 implements these values of K and B. The first stage provides an amplification of -0.72 due to the use of the inverting amplifier configuration. In the second stage, a summing amplifier is used to add the output of the first stage with a bias of 1.8 VDC. Since this stage also introduces a minus sign to the result, the overall result of a gain of 0.72 and a bias of +1.8 VDC is achieved.

Low voltage operational amplifiers, operating in the ± 2.7 to ± 5 VDC range, are readily available from Texas Instruments.

3.4   OUTPUT DEVICES

An external device should not be connected to a processor without first performing careful interface analysis to ensure the voltage, current, and timing requirements of the processor and the external device are met. In this section, we describe interface considerations for a wide variety of external devices. We begin with the interface for a single light emitting diode.

3.4.1   LIGHT EMITTING DIODES (LEDS)

LEDs may be used to indicate the logic state at a specific pin on a processor. Most LEDs have two leads: the anode or positive lead and the cathode or negative lead. When taking a "bird's eye" view of a round LED from above, one side has a slight flattening. The cathode is the lead nearest the flat portion. Also, you can hold an LED up to a light source and distinguish the cathode from the anode by its characteristic shape as shown in Figure 3.15(b).

To properly bias an LED, the anode lead must be biased at a level approximately 1.7 to 2.2 volts higher than the cathode lead. This specification is known as the forward voltage (V_f) of the LED. The LED current must also be limited to a safe current level known as the forward current (I_f). The forward diode voltage and current specifications are usually provided by the manufacturer.

A processor normally represents a logic one with a logic high voltage. In the processor documentation this is referred to as V_{OH} or the voltage when an output pin is at logic high. When at a logic high, a processor pin delivers (sources) current to the external circuit connected to it. The pin

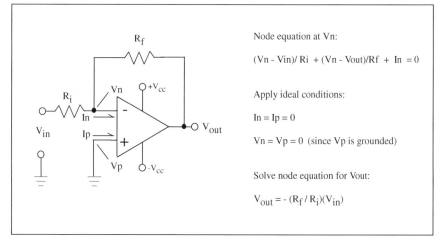

Figure 3.13: Operational amplifier analysis for the non–inverting amplifier. Adapted from [Faulkenberry].

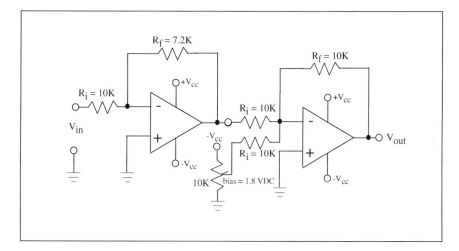

Figure 3.14: Operational amplifier implementation of the transducer interface design (TID) example circuit.

acts as a small DC power supply with an output voltage of V_{OH} with a maximum current rating of I_{OH}.

To properly interface external components to the processor, the V_{OH} and I_{OH} values of a specific pin must be determined. In this example, we interface a LED to BeagleBone header P8 pin 13. It is important to note BeagleBone documentation uses different pin names than the host

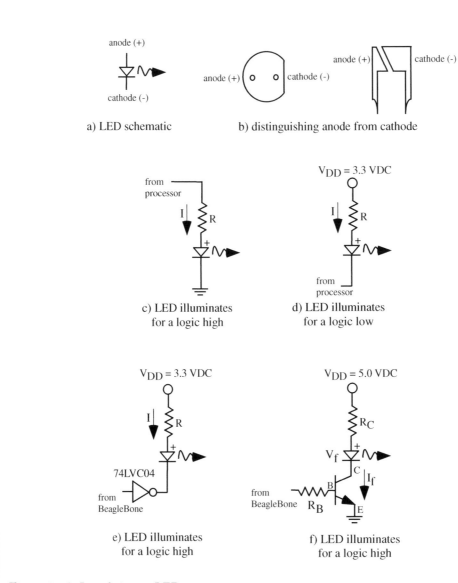

Figure 3.15: Interfacing an LED.

processor aboard. To resolve the names and find the V_{OH} and I_{OH} values for a pin, the following steps are required:

- In the BeagleBone Rev A6 System Reference Manual [Coley], determine the processor ball associated with the header pin. This information is provided on Table 8. Expansion Header P8 Pinout for pin 13. The signal name associated with pin 13 is "EHRPWM2B" and the associated processor pin is "T10."

- Looking at the processor datasheet for the pin will provide us with additional information useful for interfacing to the pin. Recall BeagleBone uses the AM3358/9 ARM Cortex–A8 microprocessor. BeagleBone uses the ZCZ package. The datasheet for this processor may be downloaded from the Texas Instruments website. Table 2–7. (Ball Characteristics) of the AM3358/9 datasheet provides the initial details. The datasheet uses the pin name "GPMC_AD9" and it is associated with ZCE and ZCZ ball pins W16 and T10, respectively. Notice that each pin has multiple functions as denoted by the mode number. This is covered in a later chapter.

- As a final step the V_{OH} and I_{OH} values for the pin may be located in Table. 3–11 of the datasheet. DC Electrical Characteristics. Note the pin names used in the table are for multiplexer mode 0. Table 3–11 is then scanned from the beginning looking for a specific pin name. If it is not found the last page of the table is used to obtain the V_{OH} and I_{OH} values. In this example, the last page of the Table is used to determine the values. For BeagleBone the $VDDSHVx = 3.3\ V$ portion of the table is used. We find the value of V_{OH} is specified as $VDDSHVx - 0.2\ V$ or 3.1 volts. The value of I_{OH} is given as 6 mA.

With the values of V_{OH} and I_{OH} determined, an interface circuit can be designed.

An LED biasing circuit is provided in Figure 3.15(c). In Figure 3.15(c) a logic one provided by the processor provides the voltage to forward bias the LED. The processor also acts as the source for the forward current through the LED. To properly bias the LED the value of the limit resistor (R) is chosen.

Example: A red (635 nm) LED is rated at 1.8 VDC with a forward operating current of 6 mA. Design a proper bias for the LED using the configuration of 3.15(c).

Answer: In the configuration of 3.15(c) the processor pin can be viewed as an unregulated power supply. That is, the pin's output voltage is determined by the current supplied by the pin. The current flows out of the processor pin through the LED and resistor combination to ground (current source). The value of R may be calculated using Ohm's Law. The voltage drop across the resistor is the difference between the 3.1 VDC supplied by the processor pin and the LED forward voltage of 1.8 VDC. The current flowing through the resistor is the LED's forward current (6 mA). This renders a resistance value of approximately 220 Ohms.

For the LED interface provided in Figure 3.15(d), the LED is illuminated when the processor provides a logic low. In this case, the current flows from the power supply back into the processor pin (current sink).

If LEDs with higher forward voltages and currents are used, alternative interface circuits may be employed. Figures 3.15(e) and (f) provide two more LED interface circuits. In Figure 3.15(e), a logic one is provided by BeagleBone to the input of the inverter. The inverter provides a logic zero at its output, which provides a virtual ground at the cathode of the LED. Therefore, the proper voltage biasing for the LED is provided. The resistor (R) limits the current through the LED. A proper resistor value can be calculated using $R = (V_{DD} - V_{DIODE})/I_{DIODE}$. It is important to note that the inverter used must have sufficient current sink capability (I_{OL}) to safely handle the forward current requirements of the LED. As in previous examples, the characteristic curves of the inverter must be carefully analyzed.

An NPN transistor such as a 2N2222 (PN2222 or MPQ2222) may be used in place of the inverter as shown in Figure 3.15(f). In this configuration, the transistor is used as a switch. When a logic low is provided by the processor, the transistor is in the cutoff region. When a logic one is provided by the microcontroller, the transistor is driven into the saturation region. To properly interface the processor to the LED, resistor values R_B and R_C must be chosen. The resistor R_B is chosen to limit the base current.

Example: Using the interface configuration of Figure 3.15(f), design an interface for an LED with V_f of 2.2 VDC and I_f of 20 mA.

Answer: In this example, we use a BeagleBone digital output pin with an I_{OH} value of 4 mA and a V_{OH} value 2.4 VDC. A loop equation, which includes these parameters, may be written as:

$$V_{OH} = (I_B \times R_B) + V_{BE}$$

Also, transistor collector current (I_c) is related to transistor base current (I_b) by the value of beta (b).

$$I_c = If = b \times Ib$$

The transistor V_{BE} is typically 0.7 VDC. Therefore, all parameters are known except R_B. Solving for R_B yields a value of 8500 Ohms. We use a value of 10 k Ohms. In this interface configuration, resistor R_C is chosen as in previous examples to safely limit the forward LED current to prescribed values. A loop equation may be written that includes R_C:

$$V_{CC} - (I_f \times R_C) - V_f - V_{CE(sat)} = 0$$

A typical value for $V_{CE(sat)}$ is 0.2 VDC. All equation values are now known except R_C. The equation may be solved yielding an R_C value of 130 Ohms.

3.4.2   SEVEN SEGMENT LED DISPLAYS

To display numeric data, seven segment LED displays are available as shown in Figure 3.16(b). Different numerals can be displayed by asserting the proper LED segments. For example, to display the number five, segments a, c, d, f, and g would be illuminated. See Figure 3.16(a). Seven segment displays are available in common cathode (CC) and common anode (CA) configurations. As the CC designation implies, all seven individual LED cathodes on the display are tied together.

As shown in Figure 3.16(b), an interface circuit is required between the processor and the seven segment LED. We use a 74LVC4245A octal bus transceiver circuit to translate the 3.3 VDC output from the processor up to 5 VDC and also provide a maximum I_{OH} value of 24 mA. A limiting resistor is required for each segment to limit the current to a safe value for the LED. Conveniently, resistors are available in DIP packages of eight for this type of application.

Seven segment displays are available in multi–character panels. In this case, separate processor pins are not used to provide data to each seven segment character. Instead, a single port equivalent (8 pins) is used to provide character data. Several other pins are used to sequence through each of the characters as shown in Figure 3.16(b). An NPN (for a CC display) transistor is connected to the common cathode connection of each individual character. As the base contact of each transistor is sequentially asserted, the specific character is illuminated. If the processor sequences through the display characters at a rate greater than 30 Hz, the display will appear to be steady and will not flicker.

3.4.3 TRI–STATE LED INDICATOR

A tri–state LED indicator is shown in Figure 3.17. It may be used to provide the status of many processor pins simultaneously. The indicator bank consists of eight green and eight red LEDs. When an individual processor pin is logic high the green LED is illuminated. When at logic low, the red LED is illuminated. If the port pin is at a tri–state, high impedance state, no LED is illuminated. Tri–state logic is used to connect a number of devices to a common bus. When a digital circuit is placed in the Hi–z (high impedance) state it is electrically isolated from the bus.

The NPN/PNP transistor pair at the bottom of the figure provides a 2.5 VDC voltage reference for the LEDs. When a specific processor pin is logic high, the green LED will be forward biased since its anode will be at a higher potential than its cathode. The 47 Ohm resistor limits current to a safe value for the LED. Conversely, when a specific processor pin is at a logic low (0 VDC), the red LED will be forward biased and illuminate. For clarity, the red and green LEDs are shown as being separate devices. LEDs are available that have both LEDs in the same device. The 74LVC4245A octal bus transceiver translates the output voltage of the processor from 3.3 VDC to 5.0 VDC.

3.4.4 DOT MATRIX DISPLAY

The dot matrix display consists of a large number of LEDs configured in a single package. A typical 5 x 7 LED arrangement is a matrix of five columns of LEDs with seven LEDs per row as shown in Figure 3.18. Display data for a single matrix column [R6–R0] is provided by the processor. That specific row is then asserted by the processor using the column select lines [C2–C0]. The entire display is sequentially illuminated a column at a time. If the processor sequences through each column fast enough (greater than 30 Hz), the matrix display appears to be stationary to a viewer.

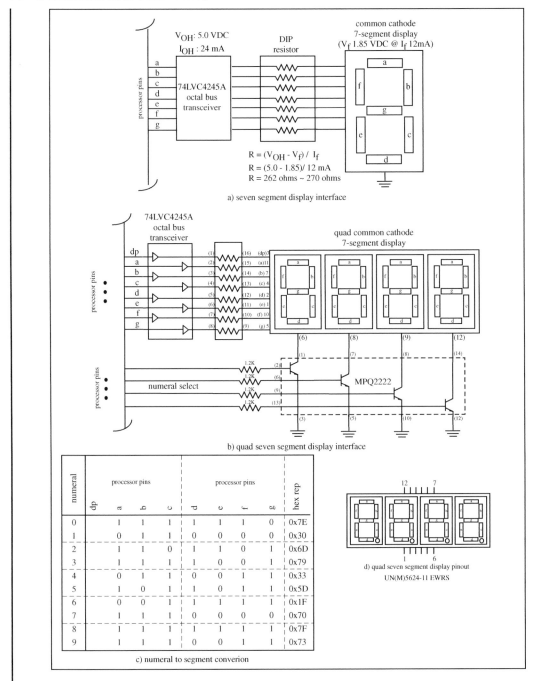

Figure 3.16: LED display devices.

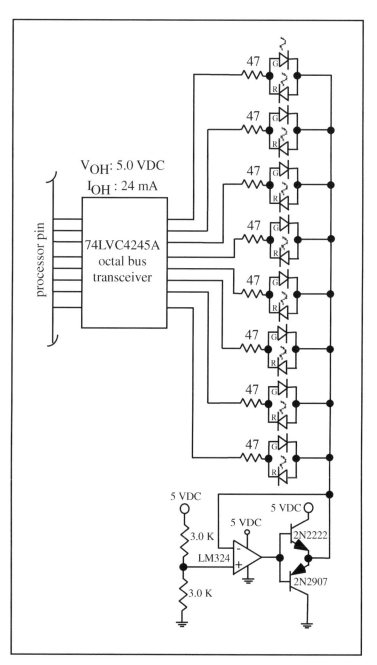

Figure 3.17: Tri-state LED display.

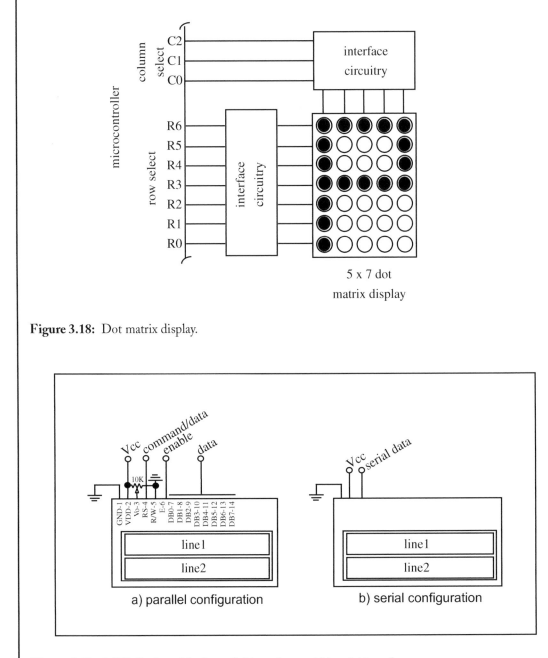

Figure 3.18: Dot matrix display.

Figure 3.19: LCD display with a) parallel interface and b) serial interface.

In Figure 3.18, we have provided the basic configuration for the dot matrix display for a single display device. However, this basic idea can be expanded in both dimensions to provide a multi–character, multi–line display. A larger display does not require a significant number of processor pins for the interface. The dot matrix display may be used to display alphanumeric data as well as graphics data. Several manufacturers provide 3.3 VDC compatible dot matrix displays with integrated interface and control circuitry.

3.4.5   LIQUID CRYSTAL DISPLAY (LCD)

An LCD is an output device to display text information as shown in Figure 3.19. LCDs come in a wide variety of configurations including multi–character, multi–line format. A 16 x 2 LCD format is common. That is, it has the capability of displaying two lines of 16 characters each. The characters are sent to the LCD via American Standard Code for Information Interchange (ASCII) format a single character at a time. For a parallel configured LCD, an eight bit data path and two lines are required between the processor and the LCD as shown in Figure 3.19(a). Many parallel configured LCDs may also be configured for a four bit data path this saving several processor pins. A small microcontroller mounted to the back panel of the LCD translates the ASCII data characters and control signals to properly display the characters. Several manufacturers provide 3.3 VDC compatible displays.

To conserve processor input/output pins, a serial configured LCD may be used. A serial LCD reduces the number of required processor pins for interface, from ten down to one, as shown in Figure 3.19(b). Display data and control information is sent to the LCD via an asynchronous serial communication link (8 bit, 1 stop bit, no parity, 9600 Baud). A serial configured LCD costs slightly more than a similarly configured parallel LCD.

3.5   HIGH POWER INTERFACES

Processors are frequently used to control a variety of high power AC and DC devices. In this section, we discuss interface techniques for a wide variety of these devices.

3.5.1   HIGH POWER DC DEVICES

A number of direct current devices may be controlled with an electronic switching device such as a MOSFET (metal oxide semiconductor field effect transistor). Specifically, an N–channel enhancement MOSFET may be used to switch a high current load (such as a motor) on or off using a low current control signal from a processor as shown in Figure 3.20. The low current control signal from the processor is connected to the gate of the MOSFET via a MOSFET driver. As shown in Figure 3.20, an LTC 1157 MOSFET driver is used to boost the control signal from the processor to be compatible with an IRLR024 power MOSFET. The IRLR024 is rated at 60 VDC V_{DS} and a continuous drain current I_D of 14 amps. The IRLR024 MOSFET switches the high current load on and off consistent with the control signal. In a low-side connection, the high current load is connected between the MOSFET source and ground.

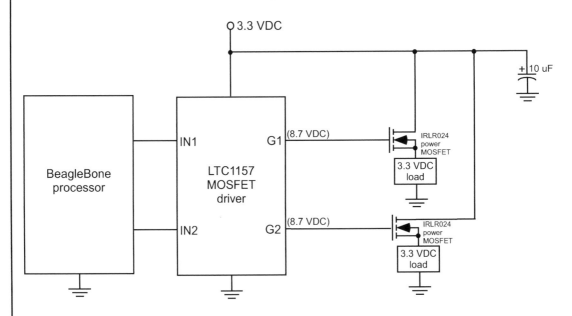

Figure 3.20: MOSFET drive circuit [adapted from Linear Technology].

3.5.2 DC MOTOR SPEED AND DIRECTION CONTROL

There are a wide variety of DC motor types that can be controlled by a processor. To properly interface a motor to the processor, we must be familiar with the different types of motor technologies. Motor types are illustrated in Figure 3.21.

General categories of DC motor types include:

- **DC motor:** A DC motor has a positive and negative terminal. When a DC power supply of suitable current rating is applied to the motor, it will rotate. If the polarity of the supply is switched with reference to the motor terminals, the motor will rotate in the opposite direction. The speed of the motor is roughly proportional to the applied voltage up to the rated voltage of the motor.

- **Servo motor:** A servo motor provides a precision angular rotation for an applied pulse width modulation (PWM) duty cycle. As the duty cycle of the applied signal is varied, the angular displacement of the motor also varies. This type of motor is used to change mechanical positions such as the steering angle of a wheel.

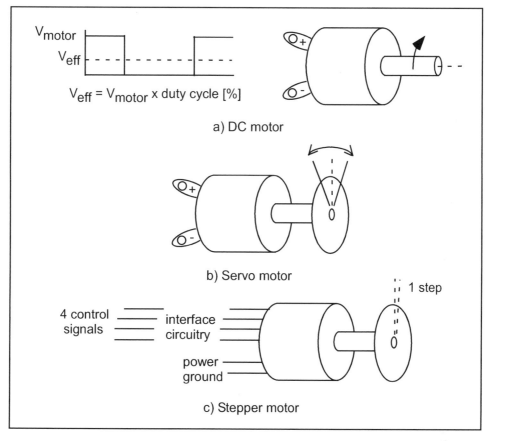

Figure 3.21: Motor types.

- **Stepper motor:** A stepper motor, as its name implies, provides an incremental step change in rotation (typically 2.5 degrees per step) for a step change in control signal sequence. Stepper motors are available with either a two or four wire interface. For the four wire stepper motor, the processor provides a four bit control sequence to rotate the motor clockwise. To turn the motor counterclockwise, the control sequence is reversed. The low power control signals are interfaced to the motor via MOSFETs or power transistors to provide for the proper voltage and current requirements of the pulse sequence. The stepper motor is used for precise positioning of mechanical components.

3.5.3 DC MOTOR OPERATING PARAMETERS

As previously mentioned, DC motor speed may be varied by changing the applied voltage. This is difficult to do with a digital control signal. However, pulse width modulation (PWM) techniques combined with a MOSFET interface circuit may be used to precisely control motor speed. The duty cycle of the PWM signal governs the percentage of the power supply voltage applied to the motor and hence the percentage of rated full speed at which the motor will rotate. The interface circuit to accomplish this type of control is shown in Figure 3.22. It is a slight variation of the control circuit provided in Figure 3.20. In this configuration, the motor supply voltage may be different than the processor's 3.3 VDC supply. For an inductive load, a reverse biased protection diode is provided across the load. The interface circuit shown allows the motor to rotate in a given direction.

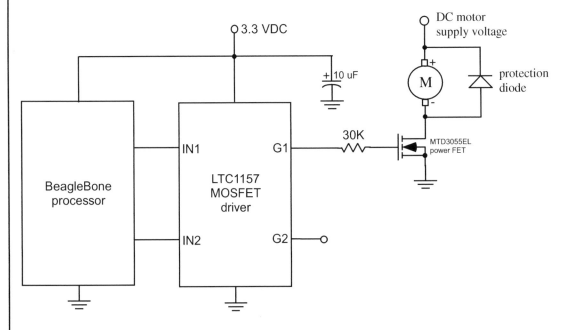

Figure 3.22: DC motor interface.

3.5.4 H–BRIDGE DIRECTION CONTROL

For a DC motor to operate in both the clockwise and counter clockwise directions, the polarity of the DC motor supplied must be changed. To operate the motor in the forward direction, the positive battery terminal must be connected to the positive motor terminal while the negative battery terminal must be attached to the negative motor terminal. To reverse the motor direction, the motor supply polarity must be reversed. An H–bridge is a circuit employed to perform this polarity switch.

An H–bridge may be constructed from discrete components as shown in Figure 3.23. The transistors Q1, Q2, Q3 and Q4 form an H–bridge. When transistors Q1 and Q4 are on, current flows from the positive terminal to the negative terminal of the motor winding. When transistors Q2 and Q3 are on, the polarity of the current is reversed, causing the motor to rotate in the opposite direction. When transistors Q3 and Q4 are on, the motor does not rotate and is in the brake state.

If PWM signals are used to drive the base of the transistors, both motor speed and direction may be controlled by the circuit. The transistors used in the circuit must have a current rating sufficient to handle the current requirements of the motor during start and stall conditions.

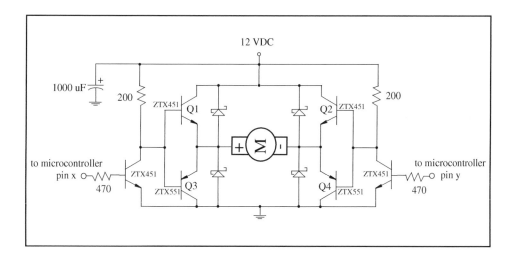

Figure 3.23: H–bridge control circuit.

Texas Instruments provides a self–contained H–bridge motor controller integrated circuit, the DRV8829. Within the DRV8829 package is a single H–bridge driver. The driver may control DC loads with supply voltages from 8 to 45 VDC with a peak current rating of 5 amps. The single H–bridge driver may be used to control a DC motor or one winding of a bipolar stepper motor [DRV8829].

3.5.5   DC SOLENOID CONTROL

The interface circuit for a DC solenoid is shown in Figure 3.24. A solenoid is used to activate a mechanical insertion (or extraction). As in previous examples, we employ the LTC1157 MOSFET driver between the processor and the power MOSFET used to activate the solenoid. A reverse biased diode is placed across the solenoid. Both the solenoid power supply and the MOSFET must have the appropriate voltage and current rating to support the solenoid requirements.

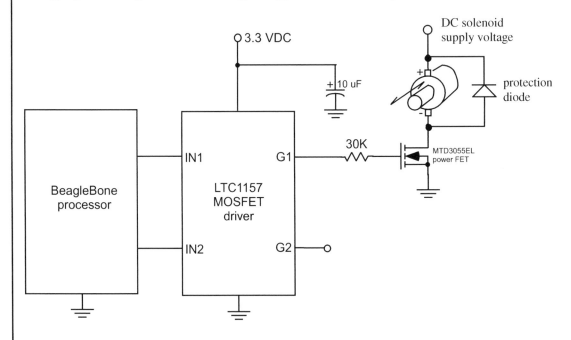

Figure 3.24: Solenoid interface circuit.

3.5.6   STEPPER MOTOR CONTROL

Stepper motors are used to provide a discrete angular displacement in response to a control signal step. There is a wide variety of stepper motors including bipolar and unipolar types with different configurations of motor coil wiring. Due to space limitations we only discuss the unipolar, 5 wire stepper motor. The internal coil configuration for this motor is shown in Figure 3.25(b).

Often, a wiring diagram is not available for the stepper motor. Based on the wiring configuration (Reference Figure 3.25(b)), one can find out the common line for both coils. It has a resistance that is one–half of all of the other coils. Once the common connection is found, one can connect the stepper motor into the interface circuit. By changing the other connections, one can determine the correct connections for the step sequence. To rotate the motor either clockwise or counter clockwise, a specific step sequence must be sent to the motor control wires as shown in Figure 3.25(b).

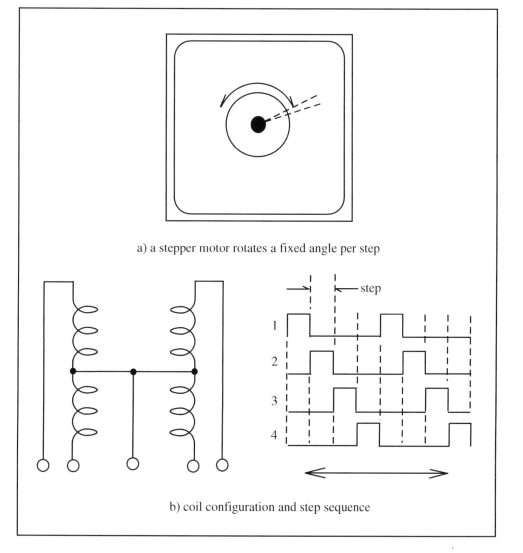

a) a stepper motor rotates a fixed angle per step

b) coil configuration and step sequence

Figure 3.25: Unipolar stepper motor.

The processor does not have sufficient capability to drive the motor directly. Therefore, an interface circuit is required as shown in Figure 3.26. The speed of motor rotation is determined by how fast the control sequence is completed. Stepper motor interfaces are available from a number of sources. An open source hardware line of stepper motor drivers, called "EasyDriver," is available from www.schmalzhaus.com/EasyDriver/.

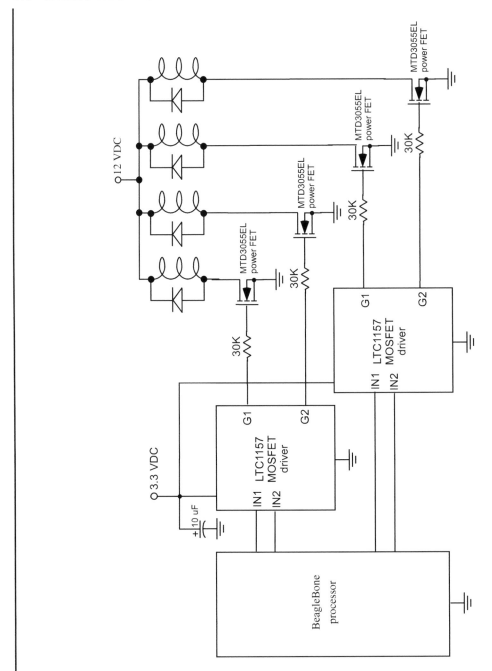

Figure 3.26: Unipolar stepper motor interface circuit.

3.6   INTERFACING TO MISCELLANEOUS DEVICES

In this section, we present a potpourri of interface circuits to connect a processor to a wide variety of peripheral devices.

3.6.1   SONALERTS, BEEPERS, BUZZERS

In Figure 3.27, we provide several circuits used to interface a processor to a buzzer, beeper or other types of annunciator devices such as a sonalert. It is important that the interface transistor and the supply voltage are matched to the requirements of the sound producing device.

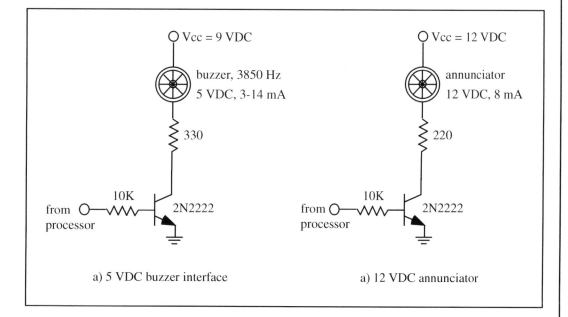

Figure 3.27: Sonalert, beepers, buzzers.

3.6.2   VIBRATING MOTOR

A vibrating motor is often used to gain one's attention as in a cell phone. These motors are typically rated at 3 VDC and a high current. The interface circuit shown in Figure 3.20 is used to drive the low voltage motor.

3.6.3   DC FAN

The interface circuit provided in Figure 3.22 may also be used to control a DC fan. As before, a reverse biased diode is placed across the DC fan motor.

3.7   AC DEVICES

A high power alternating current (AC) load may be switched on and off using a low power control signal from the processor. In this case, a Solid State Relay is used as the switching device. Solid state relays are available to switch a high power DC or AC load [Crydom]. For example, the Crydom 558–CX240D5R is a printed circuit board mounted, air cooled, single pole single throw (SPST), normally open (NO) solid state relay. It requires a DC control voltage of 3–15 VDC at 15 mA. However, this small processor compatible DC control signal is used to switch 12–280 VAC loads rated from 0.06 to 5 amps [Crydom].

To vary the direction of an AC motor, you must use a bi–directional AC motor. A bi–directional motor is equipped with three terminals: common, clockwise, and counterclockwise. To turn the motor clockwise, an AC source is applied to the common and clockwise connections. In like manner, to turn the motor counterclockwise, an AC source is applied to the common and counterclockwise connections. This may be accomplished using two of the Crydom SSRs.

3.8   APPLICATION: EQUIPPING THE BLINKY 602A ROBOT WITH A LCD

An LCD is a useful development and diagnostic tool during project development. In this section we add an LCD to the Blinky 602A robot. A low cost, 3.3 VDC, 16 character by two line LCD with a parallel connection (Sparkfun LCD–09025) is interfaced to the robot. The LCD support functions and hardware interface is provided in Figure 3.28.

There are several functions required to use the LCD including:

- LCD initialization (LCD_init): initializes LCD to startup state as specified by LCD technical data.

- LCD put character (LCD_putchar): sends a specified ASCII character to the LCD display.

- LCD print string (LCD_print): sends a string of ASCII characters to the LCD display until the null character is received.

- LCD put command (LCD_putcommand): sends a command to the LCD display.

The UML activity diagrams (UML compliant flow charts) for these functions are provided in Figure 3.28(a).

The ASCII characters and commands are sent to the LCD via an 8–bit parallel data connection and two control lines (Register Set (RS) and Enable (E)) as shown In Figure 3.28(b).

Provided below is the Bonescript code to configure the LCD and the code for the support functions.

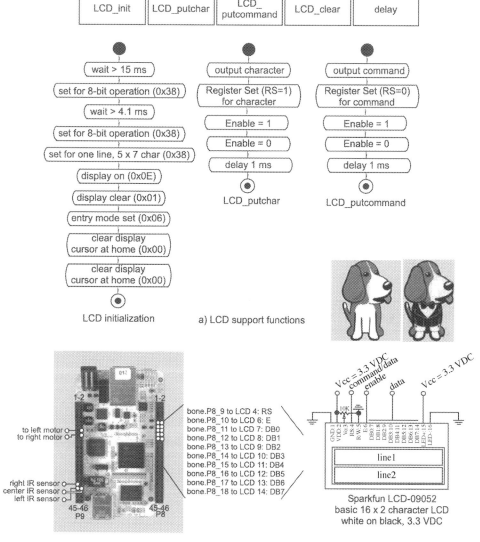

a) LCD support functions

b) LCD hardware interface

Figure 3.28: LCD for Blinky 602A robot.

```
var b = require ('bonescript');

//Old bonescript defines 'bone' globally
var pins = (typeof bone != 'undefined') ? bone : b.bone.pins;

var LCD_RS  = pins.P8_9;                //LCD Register Set (RS) control
var LCD_E   = pins.P8_10;               //LCD Enable (E) control
var LCD_DB0 = pins.P8_11;               //LCD Data line DB0
var LCD_DB1 = pins.P8_12;               //LCD Data line DB1
var LCD_DB2 = pins.P8_13;               //LCD Data line DB2
var LCD_DB3 = pins.P8_14;               //LCD Data line DB3
var LCD_DB4 = pins.P8_15;               //LCD Data line DB4
var LCD_DB5 = pins.P8_16;               //LCD Data line DB5
var LCD_DB6 = pins.P8_17;               //LCD Data line DB6
var LCD_DB7 = pins.P8_18;               //LCD Data line DB7

var counter = 1;                        //counter to be displayed

b.pinMode(LCD_RS,  b.OUTPUT);       //set pin to digital output
b.pinMode(LCD_E,   b.OUTPUT);       //set pin to digital output
b.pinMode(LCD_DB0, b.OUTPUT);       //set pin to digital output
b.pinMode(LCD_DB1, b.OUTPUT);       //set pin to digital output
b.pinMode(LCD_DB2, b.OUTPUT);       //set pin to digital output
b.pinMode(LCD_DB3, b.OUTPUT);       //set pin to digital output
b.pinMode(LCD_DB4, b.OUTPUT);       //set pin to digital output
b.pinMode(LCD_DB5, b.OUTPUT);       //set pin to digital output
b.pinMode(LCD_DB6, b.OUTPUT);       //set pin to digital output
b.pinMode(LCD_DB7, b.OUTPUT);       //set pin to digital output
LCD_init(LCD_update);               //call LCD initialize

//*****************************************************************
//LCD_update
//*****************************************************************

function LCD_update()
{
LCD_putchar(counter, next);         //write 'counter' value to LCD
```

```
//When LCD_putchar completes, schedule the next run of it
function next()
  {
  counter++;                         //update counter
  if(counter > 9)                    //Re-init after 9
    {
    counter = 1;
    LCD_init(LCD_update);
    }
  else
    setTimeout(LCD_update, 500);     //update again in 500 ms
  }
}


//*********************************************************************
//LCD_init
//*********************************************************************

function LCD_init(callback)
{
//LCD Enable (E) pin low
b.digitalWrite(LCD_E, b.LOW);

//Start at the beginning of the list of steps to perform
var i = 0;

//List of steps to perform
var steps =
  [
  function(){ setTimeout(next, 15); },          //delay 15ms
  function(){ LCD_putcommand(0x38, next); },    //set for 8-bit operation
  function(){ setTimeout(next, 5); },           //delay 5ms
  function(){ LCD_putcommand(0x38, next); },    //set for 8-bit operation
  function(){ LCD_putcommand(0x38, next); },    //set for 5 x 7 character
  function(){ LCD_putcommand(0x0E, next); },    //display on
  function(){ LCD_putcommand(0x01, next); },    //display clear
  function(){ LCD_putcommand(0x06, next); },    //entry mode set
  function(){ LCD_putcommand(0x00, next); },    //clear display, cursor home
  function(){ LCD_putcommand(0x00, callback); } //clear display, cursor home
```

```
  ];

next(); //Execute the first step

//Function for executing the next step
function next()
  {
  i++;
  steps[i-1]();
  }
}


//*******************************************************************
//LCD_putcommand
//*******************************************************************

function LCD_putcommand(cmd, callback)
{
//parse command variable into individual bits for output
//to LCD
if((cmd & 0x0080)== 0x0080) b.digitalWrite(LCD_DB7, b.HIGH);
  else b.digitalWrite(LCD_DB7, b.LOW);
if((cmd & 0x0040)== 0x0040) b.digitalWrite(LCD_DB6, b.HIGH);
  else b.digitalWrite(LCD_DB6, b.LOW);
if((cmd & 0x0020)== 0x0020) b.digitalWrite(LCD_DB5, b.HIGH);
  else b.digitalWrite(LCD_DB5, b.LOW);
if((cmd & 0x0010)== 0x0010) b.digitalWrite(LCD_DB4, b.HIGH);
  else b.digitalWrite(LCD_DB4, b.LOW);
if((cmd & 0x0008)== 0x0008) b.digitalWrite(LCD_DB3, b.HIGH);
  else b.digitalWrite(LCD_DB3, b.LOW);
if((cmd & 0x0004)== 0x0004) b.digitalWrite(LCD_DB2, b.HIGH);
  else b.digitalWrite(LCD_DB2, b.LOW);
if((cmd & 0x0002)== 0x0002) b.digitalWrite(LCD_DB1, b.HIGH);
  else b.digitalWrite(LCD_DB1, b.LOW);
if((cmd & 0x0001)== 0x0001) b.digitalWrite(LCD_DB0, b.HIGH);
  else b.digitalWrite(LCD_DB0, b.LOW);

//LCD Register Set (RS) to logic zero for command input
b.digitalWrite(LCD_RS, b.LOW);
```

```
//LCD Enable (E) pin high
b.digitalWrite(LCD_E, b.HIGH);

//End the write after 1ms
setTimeout(endWrite, 1);

function endWrite()
  {
  //LCD Enable (E) pin low
  b.digitalWrite(LCD_E, b.LOW);
  //delay 1ms before calling 'callback'
  setTimeout(callback, 1);
  }
}


//*****************************************************************
//LCD_putchar
//*****************************************************************

function LCD_putchar(chr1, callback)
{
//Convert chr1 variable to UNICODE (ASCII)
var chr = chr1.toString().charCodeAt(0);

//parse character variable into individual bits for output
//to LCD
if((chr & 0x0080)== 0x0080) b.digitalWrite(LCD_DB7, b.HIGH);
  else b.digitalWrite(LCD_DB7, b.LOW);
if((chr & 0x0040)== 0x0040) b.digitalWrite(LCD_DB6, b.HIGH);
  else b.digitalWrite(LCD_DB6, b.LOW);
if((chr & 0x0020)== 0x0020) b.digitalWrite(LCD_DB5, b.HIGH);
  else b.digitalWrite(LCD_DB5, b.LOW);
if((chr & 0x0010)== 0x0010) b.digitalWrite(LCD_DB4, b.HIGH);
  else b.digitalWrite(LCD_DB4, b.LOW);
if((chr & 0x0008)== 0x0008) b.digitalWrite(LCD_DB3, b.HIGH);
  else b.digitalWrite(LCD_DB3, b.LOW);
if((chr & 0x0004)== 0x0004) b.digitalWrite(LCD_DB2, b.HIGH);
  else b.digitalWrite(LCD_DB2, b.LOW);
if((chr & 0x0002)== 0x0002) b.digitalWrite(LCD_DB1, b.HIGH);
```

```
     else b.digitalWrite(LCD_DB1, b.LOW);
 if((chr & 0x0001)== 0x0001) b.digitalWrite(LCD_DB0, b.HIGH);
     else b.digitalWrite(LCD_DB0, b.LOW);

//LCD Register Set (RS) to logic one for character input
b.digitalWrite(LCD_RS, b.HIGH);
//LCD Enable (E) pin high
b.digitalWrite(LCD_E, b.HIGH);

//End the write after 1ms
setTimeout(endWrite, 1);

function endWrite()
  {
  //LCD Enable (E) pin low and call scheduleCallback when done
  b.digitalWrite(LCD_E, b.LOW);
  //delay 1ms before calling 'callback'
  setTimeout(callback, 1);
  }
}


//*****************************************************************
```

The following function allows a message to be sent to the LCD display. The function is called by indicating the LCD line to display the message on and the message.

```
//*****************************************************************

var b = require ('bonescript');

//Old bonescript defines 'bone' globally
var pins = (typeof bone != 'undefined') ? bone : b.bone.pins;

var LCD_RS  = pins.P8_9;      //LCD Register Set (RS) control
var LCD_E   = pins.P8_10;     //LCD Enable (E) control
var LCD_DB0 = pins.P8_11;     //LCD Data line DB0
var LCD_DB1 = pins.P8_12;     //LCD Data line DB1
var LCD_DB2 = pins.P8_13;     //LCD Data line DB2
var LCD_DB3 = pins.P8_14;     //LCD Data line DB3
var LCD_DB4 = pins.P8_15;     //LCD Data line DB4
var LCD_DB5 = pins.P8_16;     //LCD Data line DB5
```

```
var LCD_DB6 = pins.P8_17;    //LCD Data line DB6
var LCD_DB7 = pins.P8_18;    //LCD Data line DB7

b.pinMode(LCD_RS,  b.OUTPUT);        //set pin to digital output
b.pinMode(LCD_E,   b.OUTPUT);        //set pin to digital output
b.pinMode(LCD_DB0, b.OUTPUT);        //set pin to digital output
b.pinMode(LCD_DB1, b.OUTPUT);        //set pin to digital output
b.pinMode(LCD_DB2, b.OUTPUT);        //set pin to digital output
b.pinMode(LCD_DB3, b.OUTPUT);        //set pin to digital output
b.pinMode(LCD_DB4, b.OUTPUT);        //set pin to digital output
b.pinMode(LCD_DB5, b.OUTPUT);        //set pin to digital output
b.pinMode(LCD_DB6, b.OUTPUT);        //set pin to digital output
b.pinMode(LCD_DB7, b.OUTPUT);        //set pin to digital output
LCD_init(firstLine);                 //call LCD initialize
function firstLine()
{
LCD_print(1, "BeagleBone", nextLine);
}
function nextLine()
{
LCD_print(2, "Bonescript");
}


//******************************************************************
//LCD_print
//******************************************************************

function LCD_print(line, message, callback)
{
var i = 0;

if(line == 1)
  {
  LCD_putcommand(0x80, writeNextCharacter);//print to LCD line 1
  }
else
  {
  LCD_putcommand(0xc0, writeNextCharacter);//print to LCD line 2
  }
```

```
function writeNextCharacter()
  {
  //if we already printed the last character, stop and callback
  if(i == message.length)
    {
    if(callback) callback();
    return;
    }

  //get the next character to print
  var chr = message.substring(i, i+1);
  i++;

  //print it using LCD_putchar and come back again when done
  LCD_putchar(chr, writeNextCharacter);
  }
}

//******************************************************************
//LCD_init
//******************************************************************

function LCD_init(callback)
{
//LCD Enable (E) pin low
b.digitalWrite(LCD_E, b.LOW);

//Start at the beginning of the list of steps to perform
var i = 0;

//List of steps to perform
var steps =
  [
  function(){ setTimeout(next, 15); },        //delay 15ms
  function(){ LCD_putcommand(0x38, next); },  //set for 8-bit operation
  function(){ setTimeout(next, 5); },         //delay 5ms
  function(){ LCD_putcommand(0x38, next); },  //set for 8-bit operation
  function(){ LCD_putcommand(0x38, next); },  //set for 5 x 7 character
```

```
function(){ LCD_putcommand(0x0E, next); },     //display on
function(){ LCD_putcommand(0x01, next); },     //display clear
function(){ LCD_putcommand(0x06, next); },     //entry mode set
function(){ LCD_putcommand(0x00, next); },     //clear display, cursor home
function(){ LCD_putcommand(0x00, callback); } //clear display, cursor home
];

next(); //Execute the first step

//Function for executing the next step
function next()
  {
  i++;
  steps[i-1]();
  }
}


//*****************************************************************
//LCD_putcommand
//*****************************************************************

function LCD_putcommand(cmd, callback)
{
//parse command variable into individual bits for output
//to LCD
if((cmd & 0x0080)== 0x0080) b.digitalWrite(LCD_DB7, b.HIGH);
  else b.digitalWrite(LCD_DB7, b.LOW);
if((cmd & 0x0040)== 0x0040) b.digitalWrite(LCD_DB6, b.HIGH);
  else b.digitalWrite(LCD_DB6, b.LOW);
if((cmd & 0x0020)== 0x0020) b.digitalWrite(LCD_DB5, b.HIGH);
  else b.digitalWrite(LCD_DB5, b.LOW);
if((cmd & 0x0010)== 0x0010) b.digitalWrite(LCD_DB4, b.HIGH);
  else b.digitalWrite(LCD_DB4, b.LOW);
if((cmd & 0x0008)== 0x0008) b.digitalWrite(LCD_DB3, b.HIGH);
  else b.digitalWrite(LCD_DB3, b.LOW);
if((cmd & 0x0004)== 0x0004) b.digitalWrite(LCD_DB2, b.HIGH);
  else b.digitalWrite(LCD_DB2, b.LOW);
if((cmd & 0x0002)== 0x0002) b.digitalWrite(LCD_DB1, b.HIGH);
  else b.digitalWrite(LCD_DB1, b.LOW);
```

```
if((cmd & 0x0001)== 0x0001) b.digitalWrite(LCD_DB0, b.HIGH);
  else b.digitalWrite(LCD_DB0, b.LOW);

//LCD Register Set (RS) to logic zero for command input
b.digitalWrite(LCD_RS, b.LOW);
//LCD Enable (E) pin high
b.digitalWrite(LCD_E, b.HIGH);

//End the write after 1ms
setTimeout(endWrite, 1);

function endWrite()
  {
  //LCD Enable (E) pin low
  b.digitalWrite(LCD_E, b.LOW);
  //delay 1ms before calling 'callback'
  setTimeout(callback, 1);
  }
}


//********************************************************************
//LCD_putchar
//********************************************************************

function LCD_putchar(chr1, callback)
{
//Convert chr1 variable to UNICODE (ASCII)
var chr = chr1.toString().charCodeAt(0);

//parse character variable into individual bits for output
//to LCD
if((chr & 0x0080)== 0x0080) b.digitalWrite(LCD_DB7, b.HIGH);
  else b.digitalWrite(LCD_DB7, b.LOW);
if((chr & 0x0040)== 0x0040) b.digitalWrite(LCD_DB6, b.HIGH);
  else b.digitalWrite(LCD_DB6, b.LOW);
if((chr & 0x0020)== 0x0020) b.digitalWrite(LCD_DB5, b.HIGH);
  else b.digitalWrite(LCD_DB5, b.LOW);
if((chr & 0x0010)== 0x0010) b.digitalWrite(LCD_DB4, b.HIGH);
  else b.digitalWrite(LCD_DB4, b.LOW);
```

```
if((chr & 0x0008)== 0x0008) b.digitalWrite(LCD_DB3, b.HIGH);
  else b.digitalWrite(LCD_DB3, b.LOW);
if((chr & 0x0004)== 0x0004) b.digitalWrite(LCD_DB2, b.HIGH);
  else b.digitalWrite(LCD_DB2, b.LOW);
if((chr & 0x0002)== 0x0002) b.digitalWrite(LCD_DB1, b.HIGH);
  else b.digitalWrite(LCD_DB1, b.LOW);
if((chr & 0x0001)== 0x0001) b.digitalWrite(LCD_DB0, b.HIGH);
  else b.digitalWrite(LCD_DB0, b.LOW);

//LCD Register Set (RS) to logic one for character input
b.digitalWrite(LCD_RS, b.HIGH);
//LCD Enable (E) pin high
b.digitalWrite(LCD_E, b.HIGH);

//End the write after 1ms
setTimeout(endWrite, 1);

function endWrite()
  {
  //LCD Enable (E) pin low and call scheduleCallback when done
  b.digitalWrite(LCD_E, b.LOW);
  //delay 1ms before calling 'callback'
  setTimeout(callback, 1);
  }
}


//********************************************************************
```

This next example provides load average for running processes or waiting on input/output averaged over 1, 5 and 15 minutes. The Linux kernel provides these numbers via reading /proc/loadavg. We then use the LCD writing routines already discussed to perform the display.

```
//********************************************************************

var b = require('bonescript');
var fs = require('fs');

//Old bonescript defines 'bone' globally
var pins = (typeof bone != 'undefined') ? bone : b.bone.pins;

var LCD_RS  = pins.P8_9;              //LCD Register Set (RS) control
```

```
var LCD_E   = pins.P8_10;              //LCD Enable (E) control
var LCD_DB0 = pins.P8_11;              //LCD Data line DB0
var LCD_DB1 = pins.P8_12;              //LCD Data line DB1
var LCD_DB2 = pins.P8_13;              //LCD Data line DB2
var LCD_DB3 = pins.P8_14;              //LCD Data line DB3
var LCD_DB4 = pins.P8_15;              //LCD Data line DB4
var LCD_DB5 = pins.P8_16;              //LCD Data line DB5
var LCD_DB6 = pins.P8_17;              //LCD Data line DB6
var LCD_DB7 = pins.P8_18;              //LCD Data line DB7

b.pinMode(LCD_RS,  b.OUTPUT);          //set pin to digital output
b.pinMode(LCD_E,   b.OUTPUT);          //set pin to digital output
b.pinMode(LCD_DB0, b.OUTPUT);          //set pin to digital output
b.pinMode(LCD_DB1, b.OUTPUT);          //set pin to digital output
b.pinMode(LCD_DB2, b.OUTPUT);          //set pin to digital output
b.pinMode(LCD_DB3, b.OUTPUT);          //set pin to digital output
b.pinMode(LCD_DB4, b.OUTPUT);          //set pin to digital output
b.pinMode(LCD_DB5, b.OUTPUT);          //set pin to digital output
b.pinMode(LCD_DB6, b.OUTPUT);          //set pin to digital output
b.pinMode(LCD_DB7, b.OUTPUT);          //set pin to digital output
LCD_init(firstLine);                   //call LCD initialize
function firstLine()
{
LCD_print(1, "BeagleBone Load", doUpdate);
}
function doUpdate()
{
fs.readFile("/proc/loadavg", writeUpdate);
}
function writeUpdate(err, data)
{
LCD_print(2, data.toString().substring(0, 14), onUpdate);
}
function onUpdate()
{
setTimeout(doUpdate, 1000);
}

//*******************************************************************
```

```
//LCD_print
//********************************************************************

function LCD_print(line, message, callback)
{
var i = 0;

if(line == 1)
  {
  LCD_putcommand(0x80, writeNextCharacter);//print to LCD line 1
  }
else
  {
  LCD_putcommand(0xc0, writeNextCharacter);//print to LCD line 2
  }

function writeNextCharacter()
  {
  //if we already printed the last character, stop and callback
  if(i == message.length)
    {
    if(callback) callback();
    return;
    }

  //get the next character to print
  var chr = message.substring(i, i+1);
  i++;

  //print it using LCD_putchar and come back again when done
  LCD_putchar(chr, writeNextCharacter);
  }
}

//********************************************************************
//LCD_init
//********************************************************************

function LCD_init(callback)
```

```
{
//LCD Enable (E) pin low
b.digitalWrite(LCD_E, b.LOW);

//Start at the beginning of the list of steps to perform
var i = 0;

//List of steps to perform
var steps =
  [
  function(){ setTimeout(next, 15); },        //delay 15ms
  function(){ LCD_putcommand(0x38, next); },   //set for 8-bit operation
  function(){ setTimeout(next, 5); },          //delay 5ms
  function(){ LCD_putcommand(0x38, next); },   //set for 8-bit operation
  function(){ LCD_putcommand(0x38, next); },   //set for 5 x 7 character
  function(){ LCD_putcommand(0x0E, next); },   //display on
  function(){ LCD_putcommand(0x01, next); },   //display clear
  function(){ LCD_putcommand(0x06, next); },   //entry mode set
  function(){ LCD_putcommand(0x00, next); },   //clear display, cursor at home
  function(){ LCD_putcommand(0x00, callback); } //clear display, cursor at home
  ];

next(); //Execute the first step

//Function for executing the next step
function next()
  {
  i++;
  steps[i-1]();
  }
}

//*****************************************************************
//LCD_putcommand
//*****************************************************************

function LCD_putcommand(cmd, callback)
{
var chr = cmd;
```

```
//Start at the beginning of the list of steps to perform
var i = 0;

//List of steps to perform
var steps =
   [
function() {
//parse character variable into individual bits for output
//to LCD
if((chr & 0x0080)== 0x0080) b.digitalWrite(LCD_DB7, b.HIGH, next);
  else b.digitalWrite(LCD_DB7, b.LOW, next);
},function() {
if((chr & 0x0040)== 0x0040) b.digitalWrite(LCD_DB6, b.HIGH, next);
  else b.digitalWrite(LCD_DB6, b.LOW, next);
},function() {
if((chr & 0x0020)== 0x0020) b.digitalWrite(LCD_DB5, b.HIGH, next);
  else b.digitalWrite(LCD_DB5, b.LOW, next);
},function() {
if((chr & 0x0010)== 0x0010) b.digitalWrite(LCD_DB4, b.HIGH, next);
  else b.digitalWrite(LCD_DB4, b.LOW, next);
},function() {
if((chr & 0x0008)== 0x0008) b.digitalWrite(LCD_DB3, b.HIGH, next);
  else b.digitalWrite(LCD_DB3, b.LOW, next);
},function() {
if((chr & 0x0004)== 0x0004) b.digitalWrite(LCD_DB2, b.HIGH, next);
  else b.digitalWrite(LCD_DB2, b.LOW, next);
},function() {
if((chr & 0x0002)== 0x0002) b.digitalWrite(LCD_DB1, b.HIGH, next);
  else b.digitalWrite(LCD_DB1, b.LOW, next);
},function() {
if((chr & 0x0001)== 0x0001) b.digitalWrite(LCD_DB0, b.HIGH, next);
  else b.digitalWrite(LCD_DB0, b.LOW, next);
},function() {
//LCD Register Set (RS) to logic zero for command input
b.digitalWrite(LCD_RS, b.LOW, next);
},function() {
//LCD Enable (E) pin high
b.digitalWrite(LCD_E, b.HIGH, next);
```

```
},function() {
//End the write after 1ms
setTimeout(next, 1);
},function() {
  //LCD Enable (E) pin low and call scheduleCallback when done
  b.digitalWrite(LCD_E, b.LOW, next);
},function() {
  //delay 1ms before calling 'callback'
  setTimeout(callback, 1);
}];

next(); //Execute the first step

//Function for executing the next step
function next()
  {
  i++;
  steps[i-1]();
  }
}


//*****************************************************************
//LCD_putchar
//*****************************************************************

function LCD_putchar(chr1, callback)
{
//Convert chr1 variable to UNICODE (ASCII)
var chr = chr1.toString().charCodeAt(0);

//Start at the beginning of the list of steps to perform
var i = 0;

//List of steps to perform
var steps =
  [
function() {
//parse character variable into individual bits for output
//to LCD
```

```
if((chr & 0x0080)== 0x0080) b.digitalWrite(LCD_DB7, b.HIGH, next);
  else b.digitalWrite(LCD_DB7, b.LOW, next);
},function() {
if((chr & 0x0040)== 0x0040) b.digitalWrite(LCD_DB6, b.HIGH, next);
  else b.digitalWrite(LCD_DB6, b.LOW, next);
},function() {
if((chr & 0x0020)== 0x0020) b.digitalWrite(LCD_DB5, b.HIGH, next);
  else b.digitalWrite(LCD_DB5, b.LOW, next);
},function() {
if((chr & 0x0010)== 0x0010) b.digitalWrite(LCD_DB4, b.HIGH, next);
  else b.digitalWrite(LCD_DB4, b.LOW, next);
},function() {
if((chr & 0x0008)== 0x0008) b.digitalWrite(LCD_DB3, b.HIGH, next);
  else b.digitalWrite(LCD_DB3, b.LOW, next);
},function() {
if((chr & 0x0004)== 0x0004) b.digitalWrite(LCD_DB2, b.HIGH, next);
  else b.digitalWrite(LCD_DB2, b.LOW, next);
},function() {
if((chr & 0x0002)== 0x0002) b.digitalWrite(LCD_DB1, b.HIGH, next);
  else b.digitalWrite(LCD_DB1, b.LOW, next);
},function() {
if((chr & 0x0001)== 0x0001) b.digitalWrite(LCD_DB0, b.HIGH, next);
  else b.digitalWrite(LCD_DB0, b.LOW, next);
},function() {
//LCD Register Set (RS) to logic one for character input
b.digitalWrite(LCD_RS, b.HIGH, next);
},function() {
//LCD Enable (E) pin high
b.digitalWrite(LCD_E, b.HIGH, next);
},function() {
//End the write after 1ms
setTimeout(next, 1);
},function() {
  //LCD Enable (E) pin low and call scheduleCallback when done
  b.digitalWrite(LCD_E, b.LOW, next);
},function() {
  //delay 1ms before calling 'callback'
  setTimeout(callback, 1);
}];
```

```
next(); //Execute the first step

//Function for executing the next step
function next()
  {
  i++;
  steps[i-1]();
  }
}
//********************************************************************
```

3.9   APPLICATION: THE BLINKY 602A INTERFACE ON A CUSTOM CAPE

As mentioned earlier in the book, BeagleBone designers developed an elegant Cape system to interface hardware. A variety of BeagleBone Capes are available from circuitco, Inc. In this section, we develop a Blinky 602A Cape to host the IR sensor, motor and LCD interface. We employ the Adafruit Proto Cape Kit. We also provide a layout for the Blinky602A power regulator board.

Provided in Figure 3.29 is the layout and the connection diagram for the Adafruit Proto Cape for BeagleBone. The Proto Cape comes as a kit with edge connectors. The Cape may be equipped with stacking header connectors so several Capes may be stacked together. The layout provided for the connectors are also connected to corresponding nearby holes to allow for ease of hardware interface.

In Figure 3.31 the Adafruit Cape has been used to interface the Blinky 602A robot IR sensors and motor interface drivers. The schematic for this circuitry was provided in Chapter 2. For convenience we have provided a complete circuit diagram in Figure 3.30. Also, a Jameco general purpose prototyping board (#105100) is used for layout of the Blinky 602A power regulator. In Figure 3.32 the Blinky 602A robot has been equipped with the Capes and the power regulator board.

3.10   SUMMARY

In this chapter, we presented the voltage and current operating parameters for BeagleBone. We discussed how this information may be applied to properly design an interface for common input and output circuits. It must be emphasized a carefully and properly designed interface allows the processor to operate properly within its parameter envelope. If, due to a poor interface design, a processor is used outside its prescribed operating parameter values, spurious and incorrect logic values will result. We provided interface information for a wide range of input and output devices. We

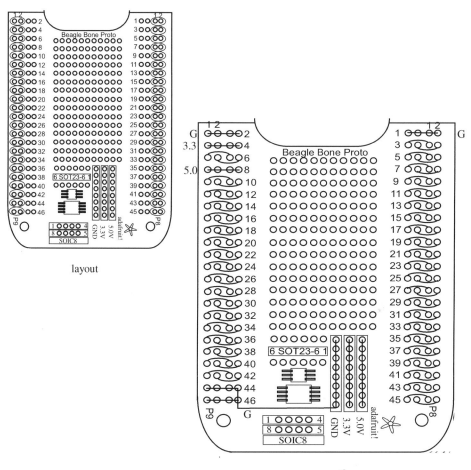

layout

connection

Figure 3.29: Adafruit BeagleBone prototype Cape.

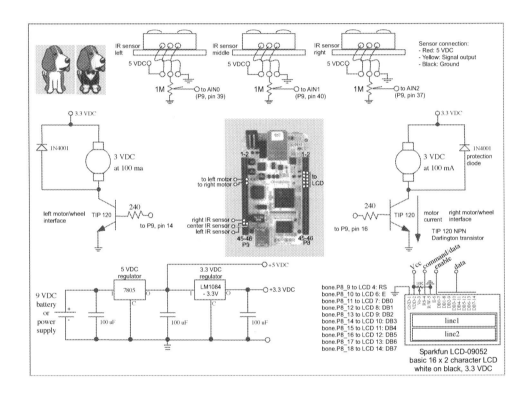

Figure 3.30: Blinky 602A interface circuit diagram.

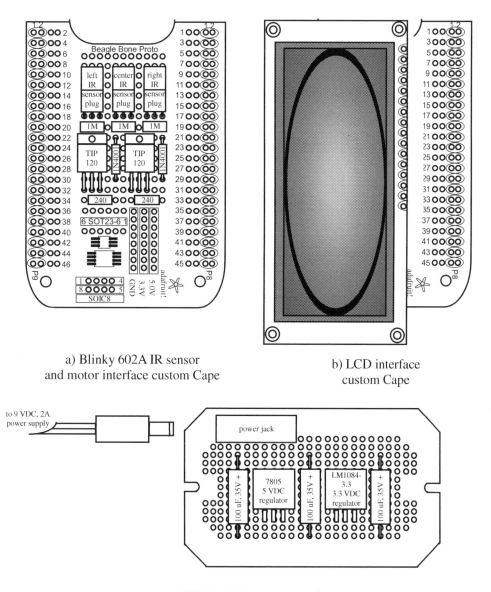

a) Blinky 602A IR sensor
and motor interface custom Cape

b) LCD interface
custom Cape

c) Blinky 602A power regulator

Figure 3.31: Blinky 602A interface.

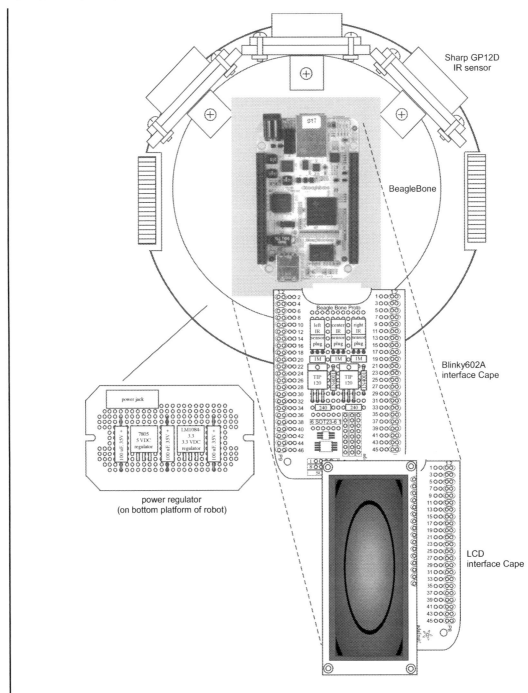

Figure 3.32: Blinky 602A with Capes and power regulator board.

also discussed the concept of interfacing a motor to the processor using PWM techniques coupled with high power MOSFET or SSR switching devices.

3.11   REFERENCES

- Mims III, Forrest M. *Getting started in Electronics.* Niles, IL: Master Publishing, 2000. Print.

- "Crydom Corporation." `www.crydom.com`

- "Sick/Stegmann Incorporated." `www.stegmann.com`

- Images Company, 39 Seneca Loop, Staten Island, NY 10314

- *Electrical Signals and Systems.* Primis Custom Publishing, McGraw–Hill Higher Education, Department of Electrical Engineering, United States Air Force Academy, CO

- Barrett, Steven and Daniel Pack. *Embedded Systems Design and Applications with the 68HC12 and HCS12.* Upper Saddle River, NJ: Pearson Prentice Hall, 2005. Print.

- Barrett, Steven and Daniel Pack. *Processors Fundamentals for Engineers and Scientists.* Morgan and Claypool Publishers, 2006. `www.morganclaypool.com`

- Barrett, Steven and Daniel Pack. *Atmel AVR Processor Primer Programming and Interfacing.* Morgan and Claypool Publishers, 2008. `www.morganclaypool.com`

- *Linear Technology, LTC1157 3.3 Dual Micropower High–Side/Low–Side MOSFET Driver.*

- *Texas Instruments H–bridge Motor Controller IC,* SLVSA74A, 2010.

- Vander Veer, Emily. *JavaScript for Dummies.* Hoboken, NJ: Wiley Publishing, Inc., 4th edition, 2005. Print.

- Pollock, John. *JavaScript.* New York, NY: McGraw Hill, 3rd edition, 2010. Print.

- Kiessling, Manuel. *The Node Beginner Guide: A Comprehensive Node.js Tutorial.* 2012. Print.

- Hughes–Croucher, Tom and Mike Wilson. *Node Up and Running.* Sebastopol, CA: O'Reilly Media, Inc., 2012. Print.

- Coley, Gerald. *BeagleBone Rev A6 Systems Reference Manual.* Revision 0.0, May 9, 2012, beagleboard.org `www.beaglebord.org`

3.12   CHAPTER EXERCISES

1. What will happen if a processor is used outside of its prescribed operating envelope?

2. Discuss the difference between the terms "sink" and "source" as related to current loading of a processor.

3. Can an LED with a series limiting resistor be directly driven by an output pin on the Beagle-Bone? Explain.

4. In your own words, provide a brief description of each of the electrical parameters of the processor.

5. What is switch bounce? Describe two techniques to minimize switch bounce.

6. Describe a method of debouncing a keypad.

7. What is the difference between an incremental encoder and an absolute encoder? Describe applications for each type.

8. What must be the current rating of the 2N2222 and 2N2907 transistors used in the tri–state LED circuit? Support your answer.

9. Draw the circuit for a six character seven segment display. Fully specify all components.

10. Repeat the problem above for a dot matrix display.

11. Repeat the problem above for an LCD display.

12. BeagleBone has been connected to a JRP 42BYG016 unipolar, 1.8 degree per step, 12 VDC at 160 mA stepper motor. The interface circuit is shown in Figure 3.26. A one second delay is used between the steps to control motor speed. Pushbutton switches SW1 and SW2 are used to assert CW and CCW stepper motion. Write the code to support this application.

13. Write the LCD print string (LCD_print_string) function. This function sends a string of ASCII characters to the LCD display until the null character is received.

CHAPTER 4

BeagleBone Systems Design

Objectives: After reading this chapter, the reader should be able to do the following:

- Define an embedded system.

- List all aspects related to the design of an embedded system.

- Provide a step–by–step approach to design an embedded system.

- Discuss design tools and practices related to embedded systems design.

- Discuss the importance of system testing.

- Apply embedded system design practices in the prototype of a BeagleBone based system with several subsystems.

- Provide a detailed design for a submersible remotely operated vehicle (ROV) including hardware layout and interface, structure chart, UML activity diagrams, and an algorithm coded in Bonescript.

- Provide a detailed design for a four wheel drive (4WD) mountain maze navigating robot including hardware layout and interface, structure chart, UML activity diagrams, and an algorithm coded in BoneScript.

4.1 OVERVIEW

In the first three chapters of the book, we introduced BeagleBone, the Bonescript programming environment, and hardware interface techniques. We pull these three topics together in this chapter. This chapter provides a step–by–step, methodical approach towards designing advanced embedded systems. In this chapter, we begin with a definition of an embedded system. We then explore the process of how to successfully (and with low stress) develop an embedded system prototype that meets established requirements. The overview of embedded system design techniques was adapted with permission from earlier Morgan and Claypool projects. We also emphasize good testing techniques. We conclude the chapter with several extended examples. The examples illustrate the embedded system design process in the development and prototype of a submersible remotely operated vehicle (ROV) and a four wheel drive (4WD) mountain maze navigating robot.

4.2   WHAT IS AN EMBEDDED SYSTEM?

An embedded system contains a processor to collect system inputs and generate system outputs. The link between system inputs and outputs is provided by a coded algorithm stored within the processor's resident memory. What makes embedded systems design so challenging and interesting is the design must also account for proper electrical interface for the input and output devices, potentially limited on–chip resources, human interface concepts, the operating environment of the system, cost analysis, related standards, and manufacturing aspects [Anderson]. Through careful application of this material you will be able to design and prototype embedded systems based on BeagleBone.

4.3   EMBEDDED SYSTEM DESIGN PROCESS

In this section, we provide a step–by–step approach to develop the first prototype of an embedded system that will meet established requirements. There are many formal design processes that we could study. We concentrate on the steps that are common to most. We purposefully avoid formal terminology of a specific approach and instead concentrate on the activities that are accomplished during the development of a system prototype. The design process we describe is illustrated in Figure 4.1 using a Unified Modeling Language (UML) activity diagram. We discuss the UML activity diagrams later in this section.

4.3.1   PROJECT DESCRIPTION

The goal of the project description step is to determine what the system is ultimately supposed to do. Questions to raise and answer during this step include but are not limited to the following:

- What is the system supposed to do?

- Where will it be operating and under what conditions?

- Are there any restrictions placed on the system design?

   To answer these questions, the designer interacts with the client to ensure clear agreement on what is to be done. The establishment of clear, definable system requirements may require considerable interaction between the designer and the client. It is essential that both parties agree on system requirements before proceeding further in the design process. The final result of this step is a detailed listing of system requirements and related specifications. If you are completing this project for yourself, you must still carefully and thoughtfully complete this step.

4.3.2   BACKGROUND RESEARCH

Once a detailed list of requirements has been established, the next step is to perform background research related to the design. In this step, the designer will ensure they understand all requirements and features required by the project. This will again involve interaction between the designer and the

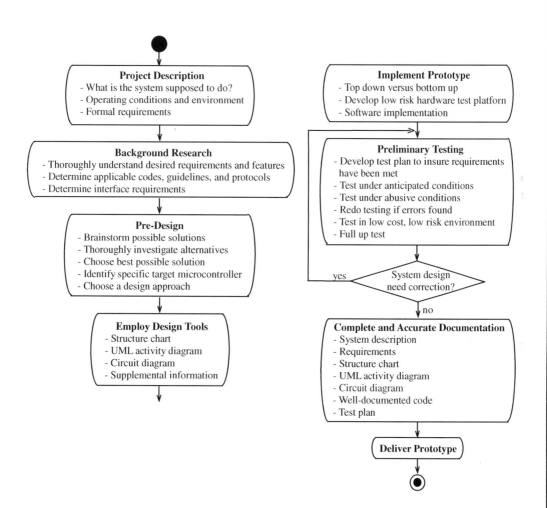

Figure 4.1: Embedded system design process.

client. The designer will also investigate applicable codes, guidelines, protocols, and standards related to the project. This is also a good time to start thinking about the interface between different portions of the input and output devices peripherally connected to the processor. The ultimate objective of this step is to have a thorough understanding of the project requirements, related project aspects, and any interface challenges within the project.

4.3.3   PRE–DESIGN

The goal of the pre–design step is to convert a thorough understanding of the project into possible design alternatives. Brainstorming is an effective tool in this step. Here, a list of alternatives is developed. Since an embedded system involves hardware and/or software, the designer can investigate whether requirements could be met with a hardware only solution or some combination of hardware and software. Generally, speaking a hardware only solution executes faster; however, the design is fixed once fielded. On the other hand, a software implementation provides flexibility but a slower execution speed. Most embedded design solutions will use a combination of both hardware and software to capitalize on the inherent advantages of each.

Once a design alternative has been selected, the general partition between hardware and software can be determined. It is also an appropriate time to select a specific hardware device to implement the prototype design. If a technology has been chosen, it is now time to select a specific processor. This is accomplished by answering the following questions:

- What processor systems or features (i.e., ADC, PWM, timer, etc.) are required by the design?

- How many input and output pins are required by the design?

- What type of memory components are required?

- What is the maximum anticipated operating speed of the processor expected to be?

Due to the variety of onboard systems, clock speed, and low cost; BeagleBone may be used in a wide array of applications typically held by microcontrollers and advanced processors.

4.3.4   DESIGN

With a clear view of system requirements and features, a general partition determined between hardware and software, and a specific processor chosen, it is now time to tackle the actual design. It is important to follow a systematic and disciplined approach to design. This will allow for low stress development of a documented design solution that meets requirements. In the design step, several tools are employed to ease the design process. They include the following:

- Employing a top–down design, bottom up implementation approach,

- Using a structure chart to assist in partitioning the system,

- Using a Unified Modeling Language (UML) activity diagram to work out program flow, and

• Developing a detailed circuit diagram of the entire system.

Let's take a closer look at each of these. The information provided here is an abbreviated version of the one provided in "Microcontrollers Fundamentals for Engineers and Scientists." The interested reader is referred there for additional details and an in–depth example [Barrett and Pack].

Top down design, bottom up implementation. An effective tool to start partitioning the design is based on the techniques of top–down design, bottom–up implementation. In this approach, you start with the overall system and begin to partition it into subsystems. At this point of the design, you are not concerned with how the design will be accomplished but how the different pieces of the project will fit together. A handy tool to use at this design stage is the structure chart. The structure chart shows how the hierarchy of system hardware and software components will interact and interface with one another. You should continue partitioning system activity until each subsystem in the structure chart has a single definable function. Directional arrows are used to indicate data flow in and out of a function.

UML Activity Diagram. Once the system has been partitioned into pieces, the next step is to work out the details of the operation of each subsystem previously identified. Rather than beginning to code each subsystem as a function, work out the information and control flow of each subsystem using another design tool: the Unified Modeling Language (UML) activity diagram. The activity diagram is simply a UML compliant flow chart. UML is a standardized method of documenting systems. The activity diagram is one of the many tools available from UML to document system design and operation. The basic symbols used in a UML activity diagram for a processor based system are provided in Figure 4.2 [Fowler].

To develop the UML activity diagram for the system, we can use a top–down, bottom–up, or a hybrid approach. In the top–down approach, we begin by modeling the overall flow of the algorithm from a high level. If we choose to use the bottom–up approach, we would begin at the bottom of the structure chart and choose a subsystem for flow modeling. The specific course of action chosen depends on project specifics. Often, a combination of both techniques, a hybrid approach, is used. You should work out all algorithm details at the UML activity diagram level prior to coding any software. If you can not explain system operation at this higher level first, you have no business being down in the detail of developing the code. Therefore, the UML activity diagram should be of sufficient detail so you can code the algorithm directly from it [Dale].

In the design step, a detailed circuit diagram of the entire system is developed. It will serve as a roadmap to implement the system. It is also a good idea at this point to investigate available design information relative to the project. This would include hardware design examples, software code examples, and application notes available from manufacturers. As before, use a subsystem approach to assemble the entire circuit. The basic building block interface circuits discussed in the previous chapter may be used to assemble the complete circuit.

At the completion of this step, the prototype design is ready for implementation and testing.

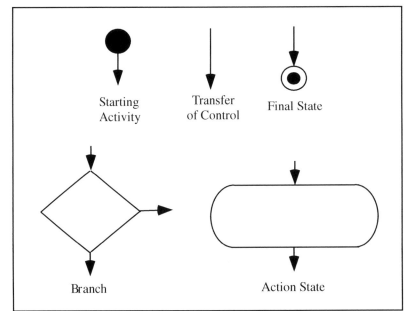

Figure 4.2: UML activity diagram symbols. Adapted from [Fowler].

4.3.5   IMPLEMENT PROTOTYPE

To successfully implement a prototype, an incremental approach should be followed. Again, the top–down design, bottom–up implementation provides a solid guide for system implementation. In an embedded system design involving both hardware and software, the hardware system including the processor should be assembled first. This provides the software the required signals to interact with. As the hardware prototype is assembled on a prototype board, each component is tested for proper operation as it is brought online. This allows the designer to pinpoint malfunctions as they occur.

Once the hardware prototype is assembled, coding may commence. As before, software should be incrementally brought online. You may use a top down, bottom up, or hybrid approach depending on the nature of the software. The important point is to bring the software online incrementally such that issues can be identified and corrected early on.

It is highly recommended that low cost stand–in components be used when testing the software with the hardware components. For example, push buttons, potentiometers, and LEDs may be used as low cost stand–in component simulators for expensive input instrumentation devices and expensive output devices such as motors. This allows you to insure the software is properly operating before using it to control the actual components.

4.3.6   PRELIMINARY TESTING

To test the system, a detailed test plan must be developed. Tests should be developed to verify that the system meets all of its requirements and also intended system performance in an operational environment. The test plan should also include scenarios in which the system is used in an unintended manner. As before, a top–down, bottom–up, or hybrid approach can be used to test the system.

Once the test plan is completed, actual testing may commence. The results of each test should be carefully documented. As you go through the test plan, you will probably uncover a number of run–time errors in your algorithm. After you correct a run–time error, the entire test plan must be repeated. This ensures that the new fix does not have an unintended effect on another part of the system. Also, as you process through the test plan, you will probably think of other tests that were not included in the original test document. These tests should be added to the test plan. As you go through testing, realize your final system is only as good as the test plan that supports it!

Once testing is complete, you should accomplish another level of testing where you intentionally try to "jam up" the system. In other words, try to get your system to fail by trying combinations of inputs that were not part of the original design. A robust system should continue to operate correctly in this type of an abusive environment. It is imperative that you design robustness into your system. When testing on a low cost simulator is complete, the entire test plan should be performed again with the actual system hardware. Once this is completed you should have a system that meets its requirements!

4.3.7   COMPLETE AND ACCURATE DOCUMENTATION

With testing complete, the system design should be thoroughly documented. Much of the documentation will have already been accomplished during system development. Documentation will include the system description, system requirements, the structure chart, the UML activity diagrams documenting program flow, the test plan, results of the test plan, system schematics, and properly documented code. To properly document code, you should carefully comment all functions describing their operation, inputs, and outputs. Also, comments should be included within the body of the function describing key portions of the code. Enough detail should be provided such that code operation is obvious. It is also extremely helpful to provide variables and functions within your code names that describe their intended use.

You might think that a comprehensive system documentation is not worth the time or effort to complete it. Complete documentation pays rich dividends when it is time to modify, repair, or update an existing system. Also, well–documented code may be often reused in other projects: a method for efficient and timely development of new systems.

In the next two sections we provide detailed examples of the system design process: a submersible robot and a four wheel drive robot capable of navigating through a mountainous maze.

4.4   SUBMERSIBLE ROBOT

The area of submersible robots is fascinating and challenging. To design a robot is quite complex. To add the additional requirement of waterproofing key components adds an additional level of challenge. (Water and electricity do not mix!) In this section we develop a control system for a remotely operated vehicle, an ROV. By definition, an ROV is equipped with a tether umbilical cable that provides power and control signals from a surface support platform. An Autonomous Underwater Vehicle (AUV) carries its own power and control equipment and does not require surface support.

We limit our discussion to the development of a BeagleBone based control system. Details on the construction and waterproofing an ROV are provided in the excellent and fascinating "Build Your Own Underwater Robot and Other Wet Projects" by Harry Bohm and Vickie Jensen. We develop a control system for the SeaPerch style of ROV as shown in Figure 4.3. There is a national–level competition for students based on the SeaPerch ROV. The goal of the program is to stimulate interest in the next generation of marine related engineering specialties [seaperch].

4.4.1   REQUIREMENTS

The requirements for this system include:

- Develop a control system to allow a three thruster (motor or bilge pump) ROV to move forward, left (port) and right (starboard).

- The ROV will be pushed down to a shallow depth via a vertical thruster and return to surface based on its own, slightly positive buoyancy.

- ROV movement will be under joystick control.

- Light emitting diodes (LEDs) are used to indicate thruster assertion.

- All power and control circuitry will be maintained in a surface support platform as shown in Figure 4.4.

4.4.2   STRUCTURE CHART

The Sea Perch structure chart is provided in Figure 4.5. As can be seen in the figure, the SeaPerch control system will accept input from the five position joystick (left, right, select, up and down). We use the Sparkfun thumb joystick (Sparkfun COM–09032) mounted to an Adafruit Proto Cape Kit for BeagleBone (Adafruit 572) as shown in Figure 4.6. The joystick schematic and connections to BeagleBone is provided in Figure 4.7.

In response to user joystick input, the SeaPerch control algorithm will issue a control command indicating desired ROV direction. In response to this desired direction command, the motor control algorithm will issue control signals to assert the appropriate motors and LEDs.

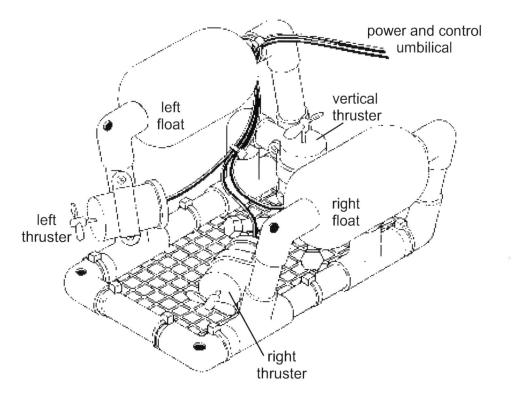

Figure 4.3: SeaPerch ROV. (Adapted and used with permission of Bohm and Jensen, West Coast Words Publishing.)

4.4.3   CIRCUIT DIAGRAM

The circuit diagram for the SeaPerch control system is provided in Figure 4.7. The thumb joystick is used to select desired ROV direction. The thumb joystick consists of two potentiometers (horizontal and vertical). A voltage of 1.8 VDC is applied to the VCC input of the joystick. As the joystick is moved, the horizontal (HORZ) and vertical (VERT) analog output voltages will change to indicate the joystick position. The joystick is also is equipped with a digital select (SEL) button. The joystick is interfaced to BeagleBone as shown in Figure 4.7.

There are three LED interface circuits connected to BeagleBone pins P8 pins 7 through 9. The LEDs illuminate to indicate the left, vertical and right thrusters have been asserted. The prime mover for the ROV will be three waterproofed motors or bilge pumps. Details on motor waterproofing may be found in "Build Your Own Underwater Robot and Other Wet Projects." The motors are driven by pulse width modulation channels (BeagleBone P9 pins 13, 14, and 16) via power FETs as shown in Figure4.7. Both the LED and the motor interfaces were discussed in the previous chapter.

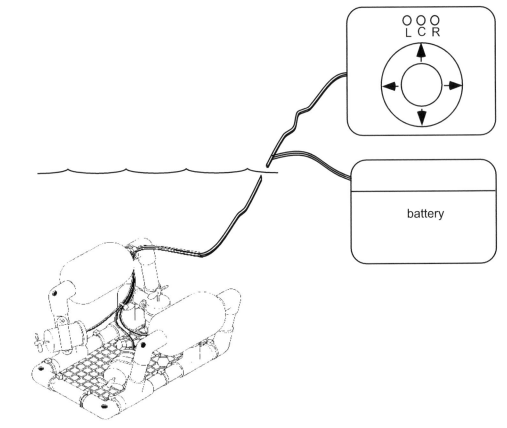

Figure 4.4: Power and control are provided remotely to the SeaPerch ROV. (Adapted and used with permission of Bohm and Jensen, West Coast Words Publishing.)

A bilge pump is a pump specifically designed to remove water from the inside of a boat. The pumps are powered from a 12 VDC source and have typical flow rates from 360 to over 3,500 gallons per minute. They range in price from US $20 to US $80.

4.4.4 UML ACTIVITY DIAGRAM

The SeaPerch control system UML activity diagram is provided in Figure 4.8. After initializing ports the control algorithm is placed in a continuous loop awaiting user input. In response to user input, the algorithm determines desired direction of ROV travel and asserts appropriate control signals for the LED and motors.

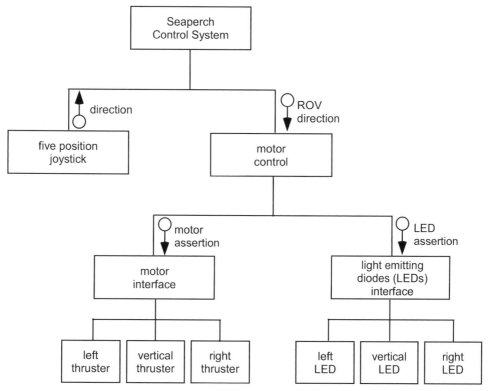

Figure 4.5: SeaPerch ROV structure chart.

4.4.5   BEAGLEBONE CODE

In this example we use the thumb joystick to control the left and right thruster (motor or bilge pump). The joystick provides a separate voltage from 0 to 1.8 VDC for the horizontal (HORZ) and vertical (VERT) position of the joystick. We use this voltage to set the duty cycle of the pulse width modulated (PWM) signals sent to the left and right thrusters. The select pushbutton (SEL) on the joystick is used to assert the vertical thruster. The relationship between the joystick position and the desired thruster activity is illustrated in Figure 4.9.

```
//**********************************************************************

var b = require('bonescript');

//Old bonescript defines 'bone' globally
var pins = (typeof bone != 'undefined') ? bone : b.bone.pins;
```

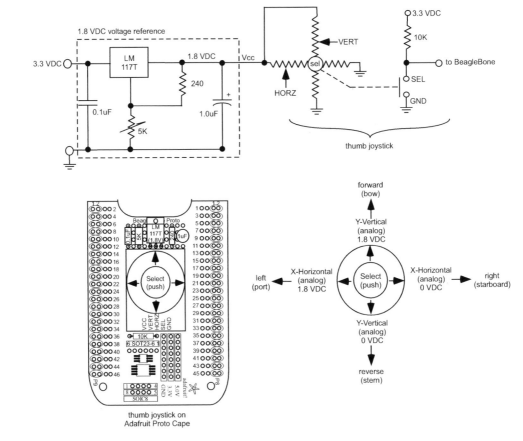

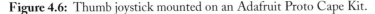

Figure 4.6: Thumb joystick mounted on an Adafruit Proto Cape Kit.

```
var joystick_horz = pins.P9_39;          //joystick horizontal signal
var joystick_vert = pins.P9_40;          //joystick vertical signal
var joystick_sel  = pins.P8_10;          //joystick select button

var left_motor_pin  = pins.P9_14;        //PWM pin for left thruster
var right_motor_pin = pins.P9_16;        //PWM pin for right thruster
var vertical_motor_pin = pins.P8_13;     //output pin for vert thruster

var left_LED_pin  = pins.P8_7;           //left LED
var vert_LED_pin  = pins.P8_8;           //vertical LED
var right_LED_pin = pins.P8_9;           //right LED
```

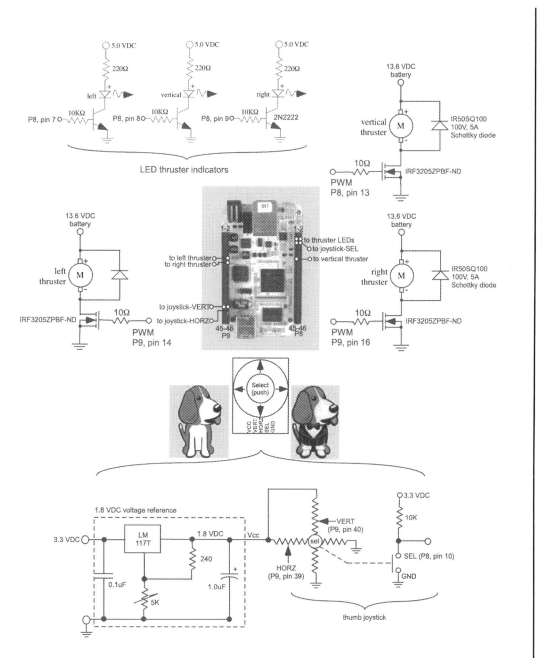

Figure 4.7: SeaPerch ROV interface control.

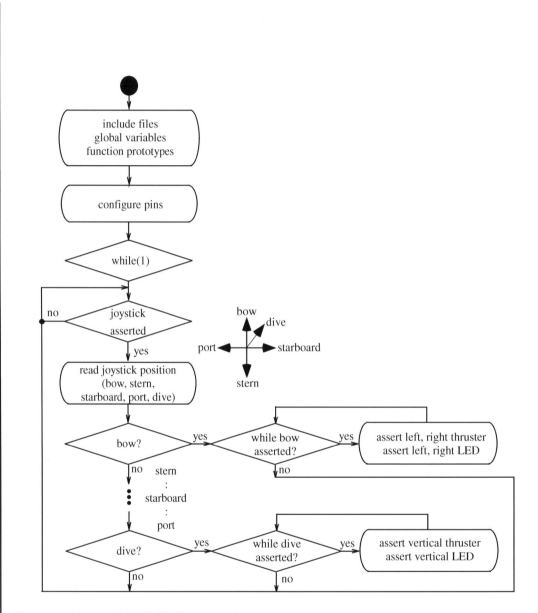

Figure 4.8: SeaPerch ROV UML activity diagram.

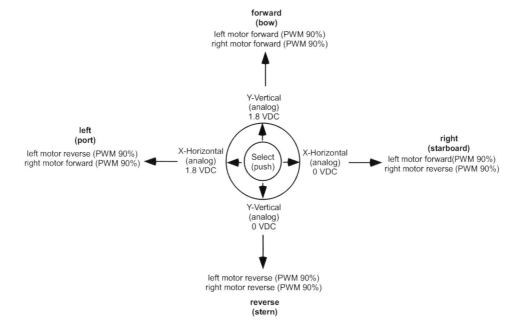

Figure 4.9: Joystick position as related to thruster activity.

```
var joystick_horizontal;
var joystick_vertical;
var vertical_thrust_on;

b.pinMode(joystick_sel, b.INPUT);
b.pinMode(vertical_motor_pin, b.OUTPUT);
b.pinMode(left_LED_pin, b.OUTPUT);
b.pinMode(vert_LED_pin, b.OUTPUT);
b.pinMode(right_LED_pin, b.OUTPUT);

//Read the joystick and react every 500 ms
setInterval(readJoystickAndDrive, 500);

function readJoystickAndDrive()
{
b.digitalWrite(vert_LED_pin, b.LOW);         //de-assert vertical LED
b.digitalWrite(left_LED_pin, b.LOW);         //de-assert left LED
b.digitalWrite(right_LED_pin, b.LOW);        //de-assert right LED
```

```
joystick_horizontal = b.analogRead(joystick_horz); //read horz position
joystick_vertical   = b.analogRead(joystick_vert); //read vert position
vertical_thrust_on  = b.digitalRead(joystick_sel); //vertical selected?

//Determine joystick position and desired thrust PWM relative
//to neutral (0.9 VDC, 0.9 VDC) position

if(joystick_horizontal > 0.9)
   joystick_horizontal = joystick_horizontal - 0.9;
else
   joystick_horizontal = 0.9 - joystick_horizontal;

if(joystick_vertical > 0.9)
   joystick_vertical = joystick_vertical - 0.9;
else
   joystick_vertical = 0.9 - joystick_vertical;

//If vertical thruster asserted - dive!
if(vertical_thrust_on)
  {
  b.digitalWrite(vertical_motor_pin, b.HIGH);   //assert vertical thruster
  b.digitalWrite(vert_LED_pin, b.HIGH);         //assert vertical LED
  }
else
  {
  b.digitalWrite(vertical_motor_pin, b.LOW);    //de-assert vert thruster
  }


//Update left and right thrusters

//Case 1:
// - Joystick horizontal has positive value
// - Joystick vertical not asserted
// - Move ROV forward
//      - Assert right and left thrusters
//      - Duty cycle determined by joystick position
if((joystick_horizontal > 0) && (joystick_vertical == 0))
```

```
   {
   b.analogWrite(left_motor_pin,  joystick_horizontal);
   b.analogWrite(right_motor_pin, joystick_horizontal);
   b.digitalWrite(left_LED_pin, b.HIGH);    //assert left LED
   b.digitalWrite(right_LED_pin, b.HIGH);   //assert right LED
   }

//insert other cases
/*
   :
   :
   :
*/
}


//****************************************************************************
```

4.4.6   PROJECT EXTENSIONS

The control system provided above has a set of very basic features. Here are some possible extensions for the system:

- Provide a powered dive and surface thruster. To provide for a powered dive and surface capability, the ROV must be equipped with a vertical thruster equipped with an H–bridge to allow for motor forward and reversal. This modification is given as an assignment at the end of the chapter.

- Left and right thruster reverse. Currently the left and right thrusters may only be powered in one direction. To provide additional maneuverability, the left and right thrusters could be equipped with an H–bridge to allow bi–directional motor control. This modification is given as an assignment at the end of the chapter.

- Proportional speed control with bi–directional motor control. Both of these advanced features may be provided by driving the H–bridge circuit with PWM signals. This modification is given as an assignment at the end of the chapter.

4.5   MOUNTAIN MAZE NAVIGATING ROBOT

In this project we extend the Blinky 602A maze navigating project described in Chapter 3 to a three–dimensional mountain pass. Also, we use a robot equipped with four motorized wheels. Each of the wheels is equipped with an H–bridge to allow bidirectional motor control. In this example we will only control two wheels. We leave the development of a 4WD robot as an end of chapter homework assignment.

4.5.1   DESCRIPTION

For this project, a DF Robot 4WD mobile platform kit was used (DFROBOT ROB0003, Jameco #2124285). The robot kit is equipped with four powered wheels. As in the Blinky 602A project, we equipped the DF Robot with three Sharp GP12D IR sensors as shown in Figure 4.10. The robot will be placed in a three-dimensional maze with reflective walls modeled after a mountain pass. The goal of the project is for the robot to detect wall placement and navigate through the maze. The robot will not be provided any information about the maze. The control algorithm for the robot is hosted on BeagleBone.

4.5.2   REQUIREMENTS

The requirements for this project are simple, the robot must autonomously navigate through the maze without touching maze walls.

4.5.3   CIRCUIT DIAGRAM

The circuit diagram for the robot is provided in Figure 4.11. The three IR sensors (left, middle, and right) are mounted on the leading edge of the robot to detect maze walls. The sensors' outputs are fed to three separate analog–to–digital (ADC) channels. The robot motors will be driven by PWM channels via an H–bridge. The robot is powered by a 7.5 VDC battery pack (5 AA batteries) which is fed to a 3.3 and 5 VDC voltage regulator. Alternatively, the robot may be powered by a 7.5 VDC power supply rated at several amps. In this case, the power is delivered to the robot by a flexible umbilical cable. The circuit diagram includes the inertial measurement unit (IMU) to measure vehicle tilt and a liquid crystal display. Both were discussed in Chapter 3.

4.5.4   STRUCTURE CHART

The structure chart for the robot project is provided in Figure 4.12.

4.5.5   UML ACTIVITY DIAGRAMS

The UML activity diagram for the robot is provided in Figure 4.13.

4.5.6   BONESCRIPT CODE

The code for the robot may be adapted from that for the Blinky602A robot. Since the motors are equipped with an H–bridge, slight modifications are required to the robot turning code. These modifications include an additional signal ($forward/\overline{reverse}$) for each H–bridge configuration to provide forward and reverse capability. For example, when forward is desired a PWM signal is delivered to one side of the H–bridge and a logic zero to the other side. A level shifter (Texas

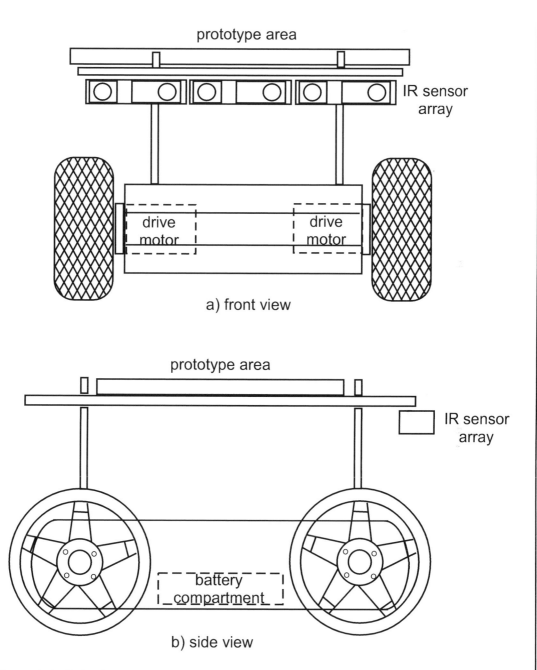

Figure 4.10: Robot layout.

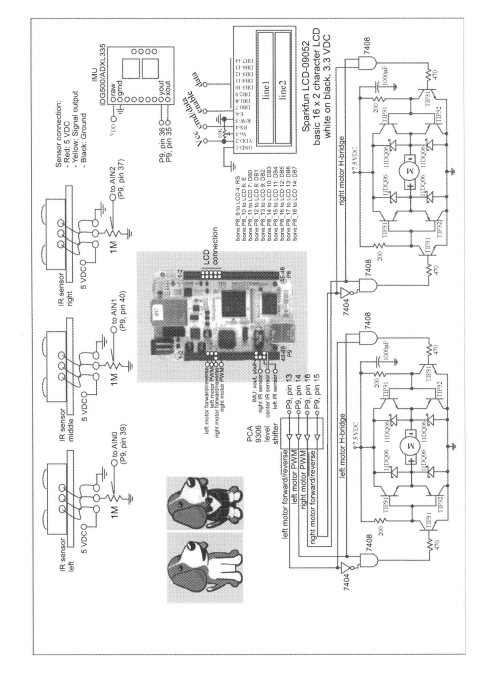

Figure 4.11: Robot circuit diagram.

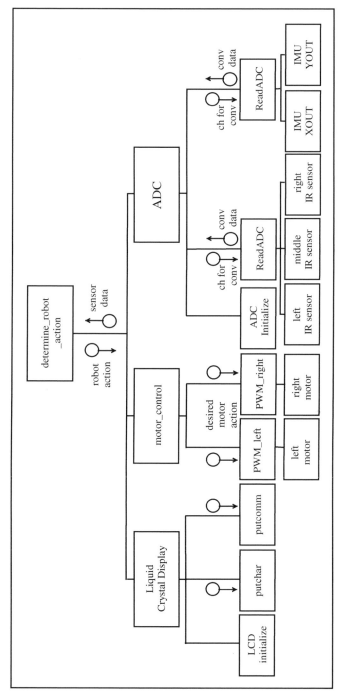

Figure 4.12: Robot structure diagram.

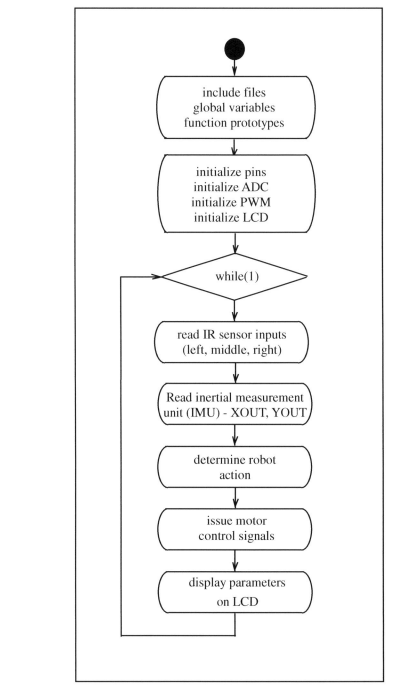

Figure 4.13: Robot UML activity diagram.

Instruments PCA9306) is used to adapt the 3.3 VDC signal output from BeagleBone to 5.0 VDC levels.

We only provide the basic framework for the Bonescript code here. Throughout Chapters 1 through 4 we have provided re–useable Bonescript code to meet system requirements.

```
//*************************************************************************
var b = require('bonescript');


//Old bonescript defines 'bone' globally
var pins = (typeof bone != 'undefined') ? bone : b.bone.pins;

var left_IR_sensor   = pins.P9_39;      //analog input for left IR sensor
var center_IR_sensor = pins.P9_40;      //analog input for center IR sensor
var right_IR_sensor  = pins.P9_37;      //analog input for right IR sensor

var IMU_xout = pins.P9_35;              //IMU xout analog signal
var IMU_yout = pins.P9_36;              //IMU yout analog signal

var left_motor_pin = pins.P9_14;        //PWM pin for left motor
var right_motor_pin = pins.P9_16;       //PWM pin for right motor
var left_motor_for_rev = pins.P9_13;    //left motor forward/reverse
var right_motor_for_rev= pins.P9_15;    //right motor forward/reverse

var left_sensor_value;
var center_sensor_value;
var right_sensor_value;


b.pinMode(left_motor_pin, b.OUTPUT);        //set pin to digital output
b.pinMode(right_motor_pin, b.OUTPUT);       //set pin to digital output
b.pinMode(left_motor_for_rev, b.OUTPUT);    //set pin to digital output
b.pinMode(right_motor_for_rev, b.OUTPUT);   //set pin to digital output

while(1)
  {
                                //read analog output from IR sensors
                                //normalized value ranges from 0..1
  left_sensor_value = b.analogRead(left_IR_sensor);
  center_sensor_value = b.analogRead(center_IR_sensor);
```

```
    right_sensor_value = b.analogRead(right_IR_sensor);

                                      //assumes desired threshold at
                                      //1.25 VDC with max value of 1.75 VDC
    if((left_sensor_value > 0.714)&&
       (center_sensor_value <= 0.714)&&
       (right_sensor_value > 0.714))
       {                                        //robot straight ahead
       b.digitalWrite(left_motor_for_rev, b.HIGH); //left motor forward
       b.analogWrite(left_motor_pin, 0.7);        //left motor RPM
       b.digitalWrite(right_motor_for_rev, b.HIGH);//right motor forward
       b.analogWrite(right_motor_pin, 0.7);       //right motor RPM
       }
/*
    else if(...)
       {
       :
       :
       :
       }
*/
    }


//*********************************************************************
```

4.5.7 MOUNTAIN MAZE

The mountain maze was constructed from plywood, chicken wire, expandable foam, plaster cloth and Bondo. A rough sketch of the desired maze path was first constructed. Care was taken to insure the pass was wide enough to accommodate the robot. The maze platform was constructed from 3/8 inch plywood on 2 by 4 inch framing material. Maze walls were also constructed from the plywood and supported with steel L brackets.

With the basic structure complete, the maze walls were covered with chicken wire. The chicken wire was secured to the plywood with staples. The chicken wire was then covered with plaster cloth (Creative Mark Artist Products #15006). To provide additional stability, expandable foam was sprayed under the chicken wire (Guardian Energy Technologies, Inc. Foam It Green 12). The mountain scene was then covered with a layer of Bondo for additional structural stability. Bondo is a two–part putty that hardens into a strong resin. Mountain pass construction steps are illustrated in 4.14. The robot is shown in the maze in Figure 4.15

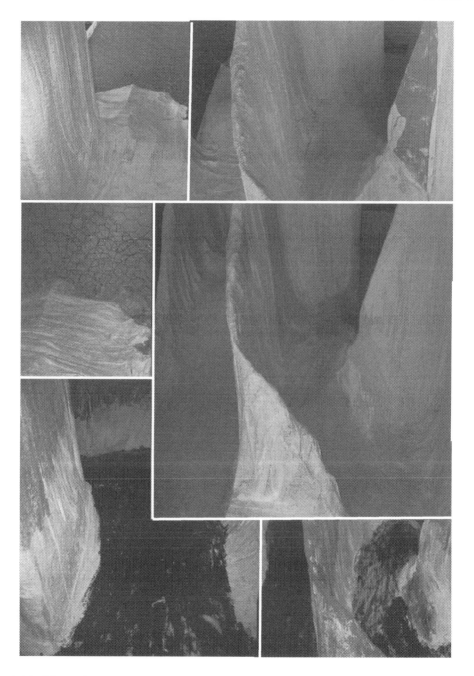

Figure 4.14: Mountain maze.

Figure 4.15: Robot in maze (photo courtesy of J. Barrett, Closer to the Sun International).

4.5.8   PROJECT EXTENSIONS

- Modify the turning commands such that the PWM duty cycle and the length of time the motors are on are sent in as variables to the function.

- Develop a function for reversing the robot.

- Equip the motor with another IR sensor that looks down toward the maze floor for "land mines." A land mine consists of a paper strip placed in the maze floor that obstructs a portion of the maze. If a land mine is detected, the robot must deactivate the maze by moving slowly back and forth for three seconds and flashing a large LED.

- The current design is a two wheel, front wheel drive system. Modify the design for a two wheel, rear wheel drive system.

- The current design is a two wheel, front wheel drive system. Modify the design for a four wheel drive system.

- Develop a four wheel drive system which includes a tilt sensor. The robot should increase motor RPM (duty cycle) for positive inclines and reduce motor RPM (duty cycle) for negatives inclines.

- Equip the robot with an analog inertial measurement unit (IMU) to measure vehicle tilt. Use the information provided by the IMU to optimize robot speed going up and down steep grades.

4.6   SUMMARY

In this chapter, we discussed the design process, related tools, and applied the process to a real world design. As previously mentioned, this design example will be periodically revisited throughout the text. It is essential to follow a systematic, disciplined approach to embedded systems design to successfully develop a prototype that meets established requirements.

4.7   REFERENCES

- Anderson, M. "Help Wanted: Embedded Engineers Why the United States is losing its edge in embedded systems." *IEEE–USA Today's Engineer*, Feb 2008.

- Barrett, Steven and Daniel Pack. *Embedded Systems Design and Applications with the 68HC12 and HCS12*. Upper Saddle River, NJ: Pearson Prentice Hall, 2005. Print.

- *Seaperch*, www.seaperch.com

- Barrett, Steven and Daniel Pack. *Processors Fundamentals for Engineers and Scientists*. Morgan and Claypool Publishers, 2006. www.morganclaypool.com

- Barrett, Steven and Daniel Pack. *Atmel AVR Processor Primer Programming and Interfacing*. Morgan and Claypool Publishers, 2008. www.morganclaypool.com

- Fowler, M. with K. Scott. *UML Distilled A Brief Guide to the Standard Object Modeling Language*. 2nd edition. Boston, MA: Addison–Wesley, 2000.

- Dale, N. and S.C. Lilly. *Pascal Plus Data Structures*. 4th edition. Englewood Cliffs, NJ: Jones and Bartlett, 1995.

4.8   CHAPTER EXERCISES

1. What is an embedded system?

2. What aspects must be considered in the design of an embedded system?

3. What is the purpose of the structure chart, UML activity diagram, and circuit diagram?

4. Why is a system design only as good as the test plan that supports it?

5. During the testing process, when an error is found and corrected, what should now be accomplished?

6. Discuss the top–down design, bottom–up implementation concept.

7. Describe the value of accurate documentation.

8. What is required to fully document an embedded systems design?

9. For the Blinky 602 robot, modify the PWM turning commands such that the PWM duty cycle and the length of time the motors are on are sent in as variables to the function.

10. For the Blinky 602 robot, equip the motor with another IR sensor that looks down for "land mines." A land mine consists of a paper strip placed in the maze floor that obstructs a portion of the maze. If a land mine is detected, the robot must deactivate it by rotating about its center axis three times and flashing a large LED while rotating.

11. For the Blinky 602 robot, develop a function for reversing the robot.

12. Provide a powered dive and surface thruster for the SeaPerch ROV. To provide for a powered dive and surface capability, the ROV must be equipped with a vertical thruster equipped with an H–bridge to allow for motor forward and reversal. This modification is given as an assignment at the end of the chapter.

13. Provide a left and right thruster reverse for the SeaPerch ROV. Currently the left and right thrusters may only be powered in one direction. To provide additional maneuverability, the left and right thrusters could be equipped with an H–bridge to allow bi–directional motor control. This modification is given as an assignment at the end of the chapter.

14. Provide proportional speed control with bi–directional motor control for the SeaPerch ROV. Both of these advanced features may be provided by driving the H–bridge circuit with PWM signals. This modification is given as an assignment at the end of the chapter.

15. For the 4WD robot, modify the PWM turning commands such that the PWM duty cycle and the length of time the motors are on are sent in as variables to the function.

16. For the 4WD robot, equip the motor with another IR sensor that looks down for "land mines." A land mine consists of a paper strip placed in the maze floor that obstructs a portion of the maze. If a land mine is detected, the robot must deactivate it by rotating about its center axis three times and flashing a large LED while rotating.

17. For the 4WD robot, develop a function for reversing the robot.

18. For the 4WD robot, the current design is a two wheel, front wheel drive system. Modify the design for a two wheel, rear wheel drive system.

19. For the 4WD robot, the current design is a two wheel, front wheel drive system. Modify the design for a four wheel drive system.

20. For the 4WD robot, develop a four wheel drive system which includes a tilt sensor. The robot should increase motor RPM (duty cycle) for positive inclines and reduce motor RPM (duty cycle) for negatives inclines.

21. Equip the robot with an inertial measurement unit (IMU) to measure vehicle tilt. Use the information provided by the IMU to optimize robot speed going up and down steep grades.

22. Develop an embedded system controlled dirigible/blimp (www.microflight.com, www.rctoys.com).

23. Develop a trip odometer for your bicycle (Hint: use a Hall Effect sensor to detect tire rotation).

24. Develop a timing system for a four lane Pinewood Derby track.

25. Develop a playing board and control system for your favorite game (Yahtzee, Connect Four, Battleship, etc.).

26. You have a very enthusiastic dog that loves to chase balls. Develop a system to launch balls for the dog.

CHAPTER 5

BeagleBone features and subsystems

Objectives: After reading this chapter, the reader should be able to do the following:

- Use the Linux operating system to interact with BeagleBone.

- Describe how the host Linux personal computer (PC) interacts with BeagleBone.

- Employ the Ångström Distribution Linux tool chain aboard BeagleBone to write, compile and execute a C and C++ program.

- Describe the features and subsystems of the ARM Cortex A8 processor.

- Describe the operation of BeagleBone exposed functions available via the P8 and P9 expansion headers.

- Program BeagleBone exposed functions using the Ångström Distribution tool chain.

5.1   OVERVIEW

In the first four chapters of this book we employed BeagleBone as a user–friendly processor. We accessed its features and subsystems via the browser–based Bonescript programming environment. The Bonescript is a powerful environment to rapidly prototype a baseline system. In the remainder of the book, we shift focus and "unleash" the power of BeagleBone as a Linux–based, 32-bit, super–scalar ARM Cortex A8 processor. We begin with a brief review of the C and C++ tool programming chain followed by examples on how to interact with the digital and analog pins aboard the processor. We then take a closer look at the features and subsystems available aboard BeagleBone. We spend a good part of the chapter describing the exposed functions of BeagleBone. These are the functions accessible to the user via the P8 and P9 extension headers. Throughout the chapter we provide sample programs on how to interact with and program the exposed functions.

5.2   PROGRAMMING BEAGLEBONE IN LINUX, C AND C++

To fully enjoy the power and capability of BeagleBone, it may be operated and programmed in the Linux environment. In this section we investigate how to use Bonescript in Linux and the C and C++ toolchain available in the Ångström Distribution Linux resident on BeagleBone.

5.2.1   BEAGLING IN LINUX

For system development BeagleBone is usually connected to a host desktop PC or laptop via a USB cable. The cable is provided with BeagleBone. An Ethernet cable is also used to connect the host with BeagleBone. It is not provided with BeagleBone but may be purchased at a local retail outlet, electronics supply store, or office supply store. The configuration forms a "mini–network" between the host computer and BeagleBone. One may wonder why a configuration like this is used. It must be emphasized that BeagleBone is a not a microcontroller–based project board. It is instead a fully functional, powerful, Linux based, expandable computer. The network configuration allows BeagleBone to share some of the useful peripheral features of the host computer during project development (e.g., keyboard, display, etc.). In Chapter 7 we describe how to implement a standalone BeagleBone based expandable computer.

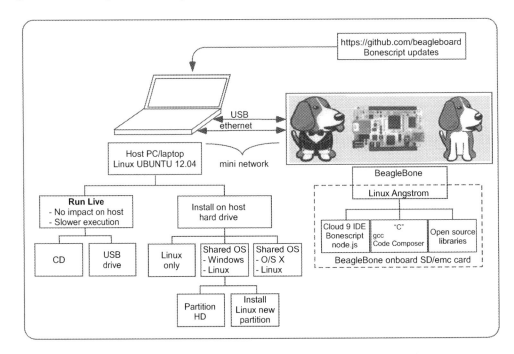

Figure 5.1: BoneScript processing environment in Linux.

It is important to note for easiest development the host PC or laptop must be equipped with the Linux operating system. Details on how to load Linux are provided in the next section. No prior use or familiarity of Linux is expected of the reader in this book. Linux is a powerful operating system originally developed by Linus Torvald of Finland in the early 1990's. Since its initial release, a variety of Linux distributions have been developed. A Linux distribution consists of the Linux operating system kernel, a collection of useful software applications and also installation software.

Linux distributions are given distinct names. As an example, we install Ubuntu Linux on the host computer while BeagleBone has the Ångström Distribution of Linux installed [Dulaney].

As shown in Figure 5.1, the host computer uses Linux release Ubuntu 12.04 (or later). There are several options on running Linux. Linux may be run in a Live mode where it is executed from a companion CD or a USB drive. In this case, the Linux operating system is not loaded on the host computer's hard drive. Alternatively, Linux may be installed on the host computer's hard drive. Linux may be installed as the only operating system for the computer or Linux may share the hard drive with another operating system such as Windows or OS X. If Linux will share the hard drive space, the hard drive must first be partitioned and space set aside for the Linux operating system. In the next section we discuss how to partition the hard drive and install Ubuntu Linux 12.04 alongside Microsoft Windows on the same PC hard drive.

BeagleBone uses the Linux Ångström Distribution. This operating system is pre–installed on the SD card on the original BeagleBone and on the eMMC Flash drive on BeagleBone Black. The Ångström Distribution includes a wide variety of useful applications. BeagleBone also arrives pre––installed with the Cloud9 IDE and Bonescript. The Bonescript is undergoing rapid development. It is essential that the most recent version of Bonescript is loaded on BeagleBone. Instructions on updating the onboard Bonescript is provided in an upcoming section. The latest software is available from www.BeagleBoard.org.

5.2.2   BEAGLEBONE LINUX RELEASES

The BeagleBone, originally released in Fall 2011, is equipped with Linux Ångström Distribution 3.2 as shown in Figure 5.2. BeagleBone Black, released in Spring 2013, is equipped with Linux Ångström Distribution 3.8. This version of Linux supports device tree overlays to configure BeagleBone hardware. As shown in Figure 5.2, the original BeagleBone may be loaded with this newer release. We discuss hardware configuration for both BeagleBone and BeagleBone Black later in the chapter.

5.2.3   BONESCRIPT PROCESSING IN LINUX

In this example, Ubuntu Linux will be loaded on a host computer that already has Microsoft Windows installed. A network is established between the host computer and BeagleBone. The steps required to establish the network include:

- Partition the hard drive on the host computer using the GParted utility provided within the Linux Ubuntu Distribution.

- Install Ubuntu Linux on the established partition.

- Configure the network between the host computer and BeagleBone.

- Start the Could9 Integrated Development Environment (IDE) resident on the BeagleBone SD card (or the eMMG flash drive).

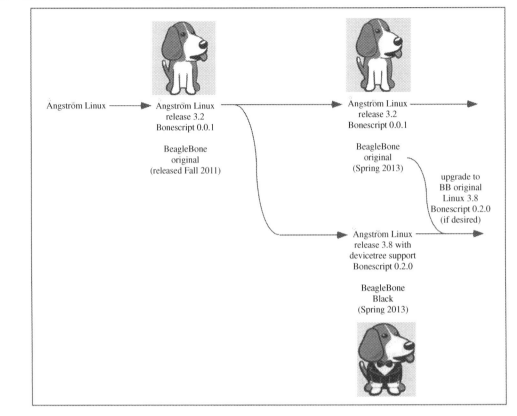

Figure 5.2: BeagleBone Linux releases.

A detailed diagram of the startup steps is provided in Figure 5.3. Additional notes on each startup step is provided for the novice user. Please reference Figure 5.3 as you reference the notes below.

- If you are a novice user, you need to find a source for Ubuntu 12.04 Linux. It is highly recommended a supplementary text on Linux is also obtained to provide additional background information on each step. A representative sample of available texts is provided in the References section of the chapter. Often a textbook companion CD/DVD will contain the operating system. The operating system may also be downloaded from www.ubuntu.com or purchased from a number of sources online.

- To partition the host computer, configure Linux to operate in the Live mode. In this mode Linux is run from a CD, DVD or USB drive and the hard drive is not modified.

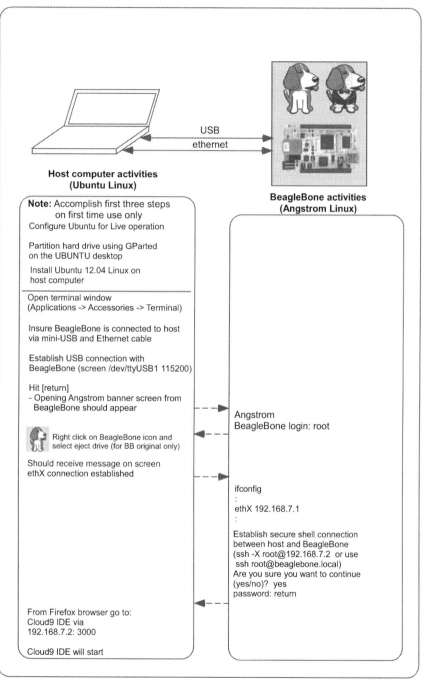

Figure 5.3: BeagleBone system startup in Linux.

- The hard drive is then partitioned using the GParted application available on the Ubuntu desktop.

- Once suitable space has been partitioned for Ubuntu Linux, Linux may be installed on the host computer hard drive.

- Every time the host computer is next started, you will be prompted for which operating system to boot. When working with BeagleBone, you may choose between Ubuntu 12.04 Linux or MS Windows.

- Once Ubuntu Linux is successfully loaded on the hard drive and started, open a terminal window using Applications − > Accessories − > Terminal.

- Insure the host computer is connected to BeagleBone via both the USB and the Ethernet cables.

- From the host computer Ubuntu terminal prompt, establish a connection with BeagleBone (screen /dev/ttyUSB1 115200).

- Hit [Return] on the host computer. The opening Ångström Linux banner should appear on the host computer screen.

- Login onto the BeagleBone operating system using the login password "root."

- At this point the host computer thinks BeagleBone is an external drive rather than a networked computer. For BeagleBone to be recognized as a networked computer, it must first be ejected as a hard drive. This is accomplished by right clicking on the Beagle icon on the host computer's Linux screen and selecting "Eject Drive."

- Once BeagleBone has been ejected as a drive, an Ethernet connection will automatically start between the host computer and BeagleBone. A message will appear on the host computer Linux screen indicating "ethX connection established." The "X" will be a number of the Ethernet connection.

- To determine the IP address of BeagleBone, the "ifconfig" command may be used on the lost computer. In response to this command, considerable information is provided including the IP address of BeagleBone.

- A secure shell connection should now be established between the host computer and Beagle-Bone (ssh − X root@beaglebone.local).

- Open the Mozilla Firefox web browser on the host computer and start Cloud9 IDE by going to 192.168.7.2:3000

A screenshot of the Cloud9 IDE is provided in Figure 5.4.

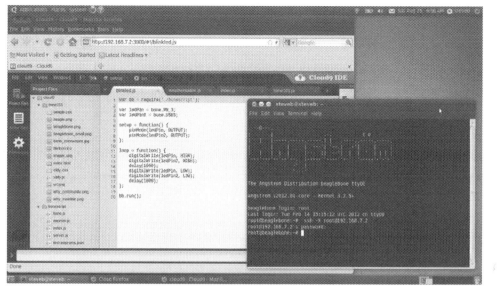

Figure 5.4: BeagleBone Cloud9 IDE screenshot.

5.3   UPDATING YOUR SD CARD OR EMMC IN LINUX

BeagleBone Bonescript and the Cloud9 IDE is a rapidly evolving, browser–based development environment. It is essential you regularly update the BeagleBone SD card (or eMMC) with the latest revision of Bonescript. To update bonescript, you need only do 'opkg update; opkg install bonescript' to get the latest version when your BeagleBone is on the Internet. Updating the image will update much more than just Bonescript, including the kernel, etc. In this section directions are provided to backup your current SD card, obtain the latest software revisions and then update the SD card. These instructions were adapted from the column "First steps with the BeagleBone [`www. borderhack.com`]."

A detailed diagram of the SD card update steps are provided in Figure 5.5. Additional notes on each update step is provided for the novice user. Please reference Figure 5.5 as you reference the notes below.

- It is important to backup the current contents of the SD card to the host computer prior to updating the SD card. This allows restoration of the SD card back to its original contents if need be. This is accomplished by making a backup directory on the host computer and changing to the host directory. To make and change to the backup directory the Linux make directory command (mkdir backup_directory_name) followed by the change directory command (cd backup_directory_name) is used.

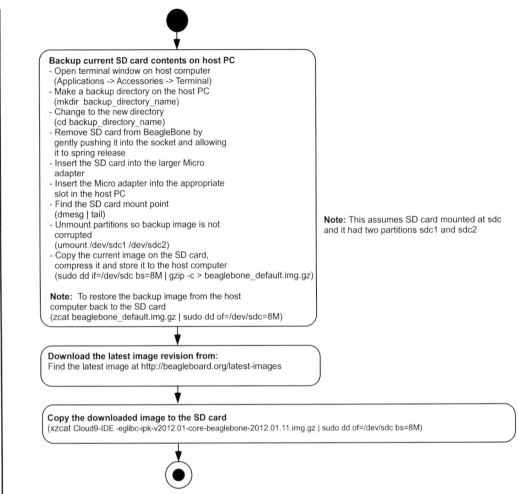

Backup current SD card contents on host PC
- Open terminal window on host computer
  (Applications -> Accessories -> Terminal)
- Make a backup directory on the host PC
  (mkdir backup_directory_name)
- Change to the new directory
  (cd backup_directory_name)
- Remove SD card from BeagleBone by
  gently pushing it into the socket and allowing
  it to spring release
- Insert the SD card into the larger Micro
  adapter
- Insert the Micro adapter into the appropriate
  slot in the host PC
- Find the SD card mount point
  (dmesg | tail)
- Unmount partitions so backup image is not
  corrupted
  (umount /dev/sdc1 /dev/sdc2)
- Copy the current image on the SD card,
  compress it and store it to the host computer
  (sudo dd if=/dev/sdc bs=8M | gzip -c > beaglebone_default.img.gz)

Note: To restore the backup image from the host
computer back to the SD card
(zcat beaglebone_default.img.gz | sudo dd of=/dev/sdc=8M)

Note: This assumes SD card mounted at sdc
and it had two partitions sdc1 and sdc2

Download the latest image revision from:
Find the latest image at http://beagleboard.org/latest-images

Copy the downloaded image to the SD card
(xzcat Cloud9-IDE -eglibc-ipk-v2012.01-core-beaglebone-2012.01.11.img.gz | sudo dd of=/dev/sdc bs=8M)

Figure 5.5: Updating the BeagleBone SD card.

- The mount location of the SD card when inserted into the host computer must then be found. This is accomplished using the Linux display message or driver message command (dmesg). This command prints the message buffer of the kernel. Since the SD card was just inserted information on this will be available. The output of the dmesg command is then routed (piped) to the Linux tail command using the "|." The tail command is used to display the last few lines of a file. The overall effect of the command sequence (dmesg | tail) is to provide information on the SD card. In this case, we are after the SD card mounting point and its partition. Look for a line that is similar to "sdc: sdc1 and sdc2." In this example the SD card is mounted at sdc and has two partitions.

- As a safety measure before proceeding the SD card is isolated from the host operating system using the Linux unmount command (umount /dev/sdc1 /dev/sdc2). The umount command detaches the SD card from the currently accessible file systems.

- The current image of the SD card is now copied, compressed and stored to the host computer (sudo dd if=/dev/sdc bs=8M | gzip $-c$ > beaglebone_default.img.gz). There is a lot going on here. The Linux substitute user do command (sudo) allows execution of a command that requires root privileges. The Linux dump device (dd) is used to copy a specified number of bytes from a specified location (in this case the SD card). The output from this command is then piped (|) to a compression algorithm via the Linux gzip command with the "$-c$" option. The "$-c$" option keeps the uncompressed version and directs it to the specified location (in this case the BeagleBone image backup file.

- Please look at `http://beagleboard.org/latest-images` for the the latest Angstrom images supported by BeagleBoard.org. The command to accomplish this is: wget `http://www.Angstrom-distribution.org/demo/beaglebone/Cloud9-IDE-eglibc-ipk-v2012.01-core-beaglebone-2012.01.11.img.gz`. The Linux wget command is used to retrieve contents from a web server. The web location is specified followed by the desired file name.

- With the latest file downloaded, the file may be uncompressed and transferred to the SD card. This is accomplished using the Linux uncompress and display data command (xzcat) and then piped to the dump device (dd) command. In this case, the destination is now the SD card.

5.3.1   PROGRAMMING IN C USING THE ÅNGSTRöM TOOLCHAIN

The Ångström Linux distribution resident on BeagleBone has built–in C and C++ programming tools. The tools of interest include the vi text editor and the gcc and g++ compilers. The vi text editor is resident within the Ångström Linux distribution onboard BeagleBone. In this section we learn how to edit, compile, and execute a small sample program.

The vi text editor is accessed from the Linux command line by typing:

>vi <filename>

For our programming example, use the following:
>vi test.cpp

Common vi commands are provided in Figure 5.6.

The next step is to install the g++ and gcc compiler on BeagleBone. This may be accomplished using the Ångström Linux command:

Table 5.1. Common vi commands.

a	insert text after cursor
A	insert text at end of current line
dd	delete current line
D	delete up to the end of current line
h or ←	move char to left
j or ↓	move down one line
k or ↑	move up one line
l or →	move char to right
yy	yank (copy) current line to buffer
nyy	yank n lines to buffer
P	put yanked line above current line
p	put yanked line below current line
x	delete character under cursor
/	finds a word going forward
:q	quit editor
:q!	quit without save
:r file	read file and insert after current line
:w file	write buffer to file
:wq	save changes and exit
Esc	end input mode, enter command mode

Figure 5.6: Common vi commands [Dulaney].

> g++ −version

This command will check to see if the latest version of this compiler is installed. If the most current version is not present, steps are provided to install it.

    Example: Use the Ångström Linux toolchain to execute a "Hello World" or in the BeagleBone case a "Aroo from BeagleBone!" The steps to accomplish this are illustrated in Figure 5.7.

5.4 BEAGLEBONE FEATURES AND SUBSYSTEMS

As discussed in Chapter 1, there are two different variants of BeagleBone: the original BeagleBone released in late 2011 and BeagleBone Black released in early 2013. In this section we provide a more

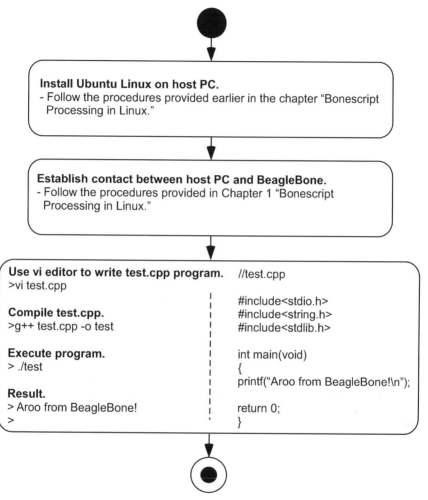

Figure 5.7: BeagleBone "Aroo!"

detailed overview of these features. The reader is encouraged to thoroughly review the "BeagleBone Rev A6 Systems Reference Manual" by Gerald Coley.

The features of both "Bones" are summarized in Figure 5.8. The two variants of BeagleBone are physically quite similar. Either fit inside a small box. The original BeagleBone is equipped with an ARM AM3359 processor that operates at a maximum frequency of 720 MHz when powered from an external 5 VDC source. The original BeagleBone is also equipped with an SD/MMC micro SD card. The card acts as the "hard drive" for BeagleBone and hosts the Ångström operating system,

Cloud9 Integrated Development Environment (IDE) and Bonescript. The original BeagleBone also features an integrated JTAG, serial port and USB cable [Coley].

BeagleBone Black also hosts the ARM AM3359 processor, but it has been updated to operate at a maximum frequency of 1 GHz when powered from an external 5 VDC source or USB. (Black may move to AM3358 in the future as the Ethercat feature isn't required for the product, but it is initially shipping with AM3359). The SD card aboard the original BeagleBone has been replaced with a 2 Giga byte (GB) eMMC (embedded MultiMedia Card) on BeagleBone Black. This provides for non–volatile mass storage in an integrated circuit package. The eMMC acts as the "hard drive" for BeagleBone Black and hosts the Ångström operating system, Cloud9 Integrated Development Environment (IDE) and Bonescript. BeagleBone has also been enhanced with a HDMI (High–Definition Multimedia Interface) framer and micro connector. HDMI is a compact digital audio/interface which is commonly used in commercial entertainment products such as DVD players, video game consoles and mobile phones. Interestingly, BeagleBone Black costs approximately half of the original BeagleBone at US $45 [Coley].

BeagleBone is equipped with a double data rate synchronous dynamic random access memory (DDR2). The DDR2 memory has a capacity of 256 Mbytes and is configured as a 128 Mbyte by 16 bit memory. BeagleBone Black updates the memory to 512MB DDR3. BeagleBone is also equipped with a 32 Kbyte memory EEPROM (electrically erasable programmable read only memory) which is a non–volatile memory used to store board configuration information [Coley].

5.5   EXPOSED FUNCTIONS

As seen in the previous section, the ARM AM3358/3359 processor hosted aboard BeagleBone has a wide variety of systems and features. Many of these features are accessible to the user via BeagleBone's expansion interface via the P8 and P9 header connectors. As mentioned earlier in the chapter, BeagleBone was originally released in Fall 2012, with Linux Ångström Distribution 3.2. BeagleBone Black, released in Spring 2013, is equipped with Linux Ångström Distribution 3.8. This version of Linux supports device tree overlays to configure BeagleBone hardware. In this section we investigate how to gain access to exposed features on the original BeagleBone via the expansion interface. In the next section we discuss how access these features for BeagleBone Black using device tree overlays.

5.5.1   EXPANSION INTERFACE – ORIGINAL BEAGLEBONE

Figure 5.9 illustrates the BeagleBone expansion interface via expansion header P9 and P8. For each pin the signal name and pin description is provided in Table 11 for header P9 and Table 8 for header P8. These tables are available in the "BeagleBone System Reference Manual (Rev A6.0.0) [Coley]." Because of the multiple features aboard the AM3358/3359 processor, each pin may have up to eight different functions per pin. The alternate functions for each pin are provided in Tables 12 and 13 of the manual for the P9 header and Tables 9 and 10 for the P8 header.

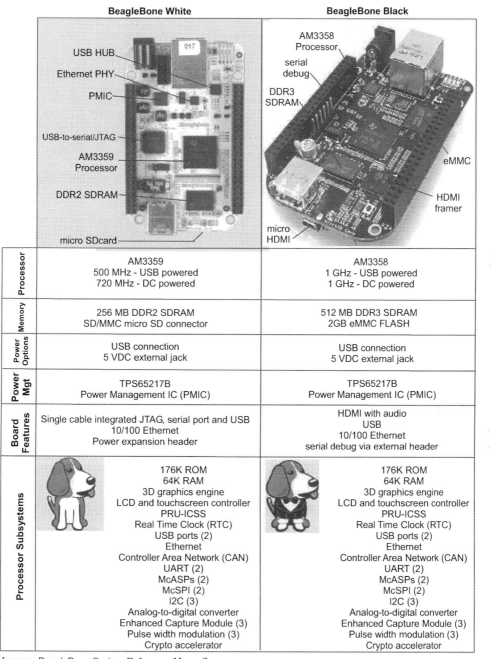

	BeagleBone White	BeagleBone Black
Processor	AM3359 500 MHz - USB powered 720 MHz - DC powered	AM3358 1 GHz - USB powered 1 GHz - DC powered
Memory	256 MB DDR2 SDRAM SD/MMC micro SD connector	512 MB DDR3 SDRAM 2GB eMMC FLASH
Power Options	USB connection 5 VDC external jack	USB connection 5 VDC external jack
Power Mgt	TPS65217B Power Management IC (PMIC)	TPS65217B Power Management IC (PMIC)
Board Features	Single cable integrated JTAG, serial port and USB 10/100 Ethernet Power expansion header	HDMI with audio USB 10/100 Ethernet serial debug via external header
Processor Subsystems	176K ROM 64K RAM 3D graphics engine LCD and touchscreen controller PRU-ICSS Real Time Clock (RTC) USB ports (2) Ethernet Controller Area Network (CAN) UART (2) McASPs (2) McSPI (2) I2C (3) Analog-to-digital converter Enhanced Capture Module (3) Pulse width modulation (3) Crypto accelerator	176K ROM 64K RAM 3D graphics engine LCD and touchscreen controller PRU-ICSS Real Time Clock (RTC) USB ports (2) Ethernet Controller Area Network (CAN) UART (2) McASPs (2) McSPI (2) I2C (3) Analog-to-digital converter Enhanced Capture Module (3) Pulse width modulation (3) Crypto accelerator

[source: BeagleBone System Reference Manual]

Figure 5.8: BeagleBone comparison [Coley]. (Figures adapted and used with permission of beagleboard.org.)

Table 9 and 10. P8 MUX options mode, 0-3, 4-7

Table 8.   Expansion Header P8 Pinout

SIGNAL NAME	PROC	CONN	CONN	PROC	SIGNAL NAME
GND	GND	1	2	GND	GND
GPIO1_6	R9	3	4	T9	GPIO1_7
GPIO1_2	R8	5	6	T8	GPIO1_3
TIMER4	R7	7	8	T7	TIMER7
TIMER5	T6	9	10	U6	TIMER6
GPIO1_13	R12	11	12	T12	GPIO1_12
GPIO1_15	T19	13	14	T11	GPIO0_26
GPIO0_27	U13	15	16	V13	GPIO1_14
EHRPWM2A	U12	17	18	V12	GPIO2_1
GPIO1_30	U10	19	20	V9	GPIO1_31
GPIO1_4	L9	21	22	V8	GPIO1_5
GPIO1_0	L8	23	24	V7	GPIO1_1
GPIO2_22	L7	25	26	V6	GPIO1_29
GPIO2_23	U5	27	28	V5	GPIO2_24
UART5_CTSN	R6	29	30	U8	GPIO2_25
UART4_RTSN	V4	31	32	T5	UART5_RTSN
UART4_CTSN	V3	33	34	U4	UART3_RTSN
UART5_TXD	V2	35	36	U3	UART3_CTSN
GPIO2_12	U1	37	38	U2	UART5_RXD
GPIO2_10	T3	39	40	T4	GPIO2_13
GPIO2_8	T1	41	42	T2	GPIO2_11
GPIO2_6	R3	43	44	R4	GPIO2_9
	R1	45	46	R2	GPIO2_7

Table 12 and 13. P9 MUX options mode, 0-3, 4-7

Table 11.   Expansion Header P9 Pinout

SIGNAL NAME	PIN	CONN	CONN	PIN	SIGNAL NAME
GND	GND	1	2	GND	
VDD_3V3EXP	VDD_3V3EXP	3	4	VDD_3V3EXP	
VDD_5V	VDD_5V	5	6	VDD_5V	
SYS_5V	SYS_5V	7	8	SYS_5V	
PWR_BUT*		9	10	A10	SYS_RESETn
UART4_RXD	T17	11	12	U18	GPIO1_28
UART4_TXD	U17	13	14	U14	EHRPWM1A
GPIO1_16	B13	15	16	T14	EHRPWM1B
I2C1_SCL	A16	17	18	B16	I2C1_SDA
I2C2_SCL	D17	19	20	D18	I2C2_SDA
UART2_TXD	B17	21	22	A17	UART2_RXD
GPIO1_17	Y14	23	24	D15	UART1_TXD
GPIO3_21	A14	25	26	D16	UART1_RXD
GPIO3_19	C13	27	28	C12	SPI1_CS0
SPI1_D0	B13	29	30	B12	SPI1_D1
SPI1_SCLK	A13	31	32	VDD_ADC(1.8V)	
AIN4	C8	33	34	GNDA_ADC	
AIN6	A5	35	36	A5	AIN5
AIN2	B6	37	38	A7	AIN3
AIN0	D14	39	40	C7	AIN1
ClkOUT2	D14	41	42	C18	GPIO0_7
	GND	43	44	GND	
	GND	45	46	GND	

Figure 5.9: BeagleBone expansion interface [Coley].

The multiple functions of a given pin are controlled by a multiplexer as shown in Figure 5.10a). The multiplexer acts as a multi–position switch. In the figure, an 8–to–1 multiplexer is shown. This multiplexer has eight inputs (I[0] through I[7]) and one output (O). Only one input may be connected to the output at any given time. The input is selected via the selected switches (S[2:0]). This has important implications in BeagleBone based systems. When designing a system, you will not be able to simultaneously use systems that share the same expansion header pins without careful precaution.

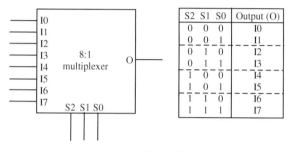

a) 8:1 multiplexer

from Table 12, 13: Connector P9 Signal Pin Mux Options [Coley]

Pin	Proc	Signal Name	MODE0	MODE1	MODE2	MODE3	MODE4	MODE5	MODE6	MODE7
42	C18	GPIO0_7	eCAP0_in_PWM0_out	uart3_txd	spi1_cs1	pr1_ecap0_ecap_...	spi1_sclk	mmc0_sdwp	xdma_event_intr2	gpio0_7

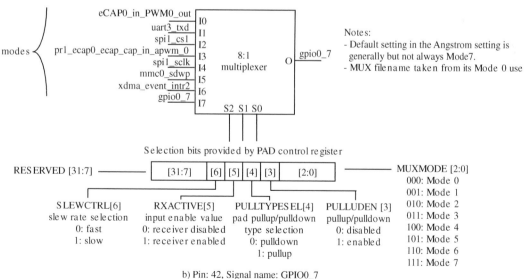

b) Pin: 42, Signal name: GPIO0_7

Figure 5.10: Processor pin [Coley].

As previously mentioned, Tables 12 and 13 for the P9 header and Tables 9 and 10 for the P8 header provide all the different modes (Mode 0 through Mode 7). The modes may be viewed as the multiplexer inputs and the processor pin as the multiplexer output. These tables also provide the:

- **PIN:** pin number on the expansion header (P9 or P8).

- **PROC:** pin number of the processor.

- **Signal name:** the signal name of the pin.

As an example, Figure 5.10b) provides a detailed illustration of the GPIO0_7 pin. An extract of Tables 12 and 13 provide the important information for the pin:

- **PIN:** 42

- **PROC:** C18

- **Signal name:** GPIO0_7

- **MODE[0] to MODE[7]:** the modes for this pin are shown as inputs to the multiplexer in Figure 5.10b).

Each processor pin has a corresponding PAD Control Register that contains the select pins for the multiplexer to select the pin mode and also other pin attributes. As shown in Figure 5.10b) the PAD control register is a 32–bit register that contains settings for:

- **Slew rate selection (SLEWCTRL):** 0: fast and 1: slow

- **Receiver input enabled (RXACTIVE):** 0: disabled and 1: enabled

- **Pad pullup type selection (PULLTYPESEL):** 0: pulldown and 1: pullup

- **Pullup/pulldown selection (PULLEDEN):** 0: disabled and 1: enabled

- **Multiplexer mode selection (MUXMODE):** 000 (Mode 0) through 111 (Mode 7)

The PAD control settings are fairly self–explanatory. Let's take a closer look at pullup and pulldown resistors. As mentioned in Chapter 3 (see Figure 3.4), a pullup or pulldown resistor is required with various switch configurations. As shown in Figure 5.11, switches may be configured in different ways to provide various logic transitions. Pullup or pulldown resistors may be activated on the processor pin to aid in switch interfacing.

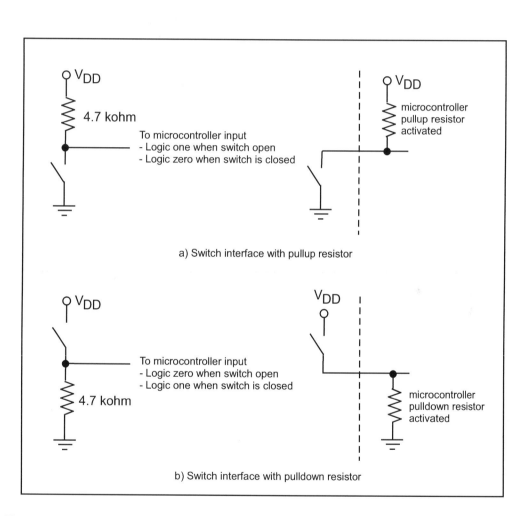

Figure 5.11: Switch interface with pullup and pull down resistors.

5.5.2 ACCESSING PINS VIA LINUX 3.2

The Ångström Linux Distribution 3.2 uses a file system to access various pins and PAD control registers associated with the AM3358/3359 processor. That is, each pin and PAD control register is accessed as though it were a file. This virtual file system is referred to as sysfs. The sysfs exports information about devices and drivers from the kernel to the user space. For the Ångström Linux operating system, the files are located in various directories. Sometimes the file locations are moved between different revisions of the operating system. Typically, files of interest may be found in the following BeagleBone directories:

```
/sys/kernel/debug/omap_mux
/sys/class/gpio
/sys/devices/platform/omap/tsc
/sys/devices/platform/tsc
```

There are several files associated with each general purpose input/output (GPIO) pin. They are:

- The PAD control file that contains the multiplexer and pin configuration information (slew rate, pullup/pulldown, multiplexer setting, etc.).

- The GPIO "direction" (input (in) or output (out)) and "value" (logic high (1) or low (0)) files.

The PAD control file is typically named after the MODE 0 function of the pin. For example, from Table 9, pin 3 on the P8 expansion header has the following specifications:

- **PROC:** R9

- **NAME:** GPIO1_6

- **MODE0:** gpmc_ad6

The PAD control file for the GPIO1_6 pin is named "gpmc_ad6." This file may be found at:

```
/sys/kernel/debug/omap_mux/gpmc_ad6
```

To determine the file to access the GPIO1_6 pin, we must translate its name to its corresponding pin number. The general form of the pin name is:

$$GPIO \ < bank \ number > \_ < pin \ number \ within \ bank >$$

The general purpose pins are divided into three banks: 0, 1, 2 and 3. Each bank has a corresponding offset:

- bank: 0, offset: 0

- bank: 1, offset: 32

- bank: 2, offset: 64

- bank: 3, offset: 96

Also, the pin's number within the bank must be taken into account. The final pin number is determined using:

$$pin\ number\ =\ (bank\ number\ *\ 32)\ +\ pin\ number\ within\ bank$$

For example, for the GPIO1_6 pin, the overall pin number is:

$$pin\ number\ =\ (1\ *\ 32)\ +\ 6\ =\ 38$$

There are two files associated with this pin: "direction" and "value." To create these files, the pin number must be sent to an export file:

```
>echo 38 > /sys/class/gpio/export
```

The two files created are then located in the gpio38 directory:

```
/sys/class/gpio/gpio38/direction
/sys/class/gpio/gpio38/value
```

Accessing the GPIO files in Linux 3.2
Since the processor pins are controlled via a file system, they can be accessed from the Linux command line using the "cat" and the "echo" commands.

The cat command is used to concatenate and display files. That is, it displays a file's contents to the computer screen. For example, to display the contents of the PAD control register for GPIO1_6, the following command may be used:

```
>cat /sys/kernel/debug/omap_mux/gpmc_ad6
```

This will result in the following display on the screen:

```
>cat /sys/kernel/debug/omap_mux/gpmc_ad6
name: gpmc_ad6.gpio1_6 (0x44e10818/0x818 = 0x0037), b NA, t NA
mode: OMAP_PIN_INPUT_PULLUP | OMAP_MUX_MODE7
signals: gpmc_ad6 | mmc1_dat6 | NA | NA | NA | NA | NA | gpio1_6
```

The echo command may be used to write a value to a file. The general format of the command is:

```
>echo <some value>  > file_name
```

For example, to set the direction of the GPIO1_6 pin, the following commands may be used:

```
>echo out > /sys/class/gpio/gpio38/direction
>echo 1 > /sys/class/gpio/gpio38/value
```

Example: Illuminate an LED connected to the GPIO1_6 pin using Linux commands.

In Chapter 1 we illuminated an LED connected to the GPIO1_6 pin using the circuit illustrated in Figure 5.12. For BeagleBone black the LED should be connected to header P8 pin 13.

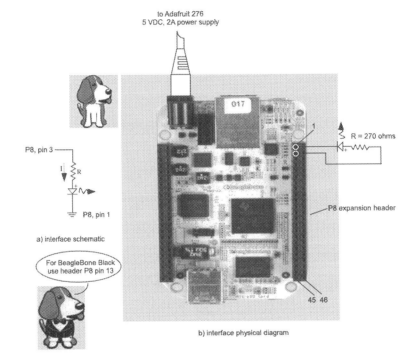

Figure 5.12: Interfacing an LED to BeagleBone. **Note:** For BeagleBone black connect the LED to header P8 pin 13.

Use the following commands to configure the GPIO1_6 pin multiplexer and turn the LED on and off.

```
Configure the GPIO1_6 pin:
- slew rate: slow (0)
- receiver: disabled (0)
- pad pullup type: NA (0)
- pullup/pulldown selection: disabled (0)
- mux mode: GPIO1_6 (7)
```

```
>echo 7 > /sys/kernel/debug/omap_mux/gpmc_ad6
```

```
Establish the GPIO1_6 direction and output files:
>echo 38 > /sys/class/gpio/export
```

```
Set GPIO1_6 for output:
>echo out > /sys/class/gpio/gpio38/direction
```

```
Set GPIO1_6 logic high (1):
>echo 1 > /sys/class/gpio/gpio38/value
```

```
Set GPIO1_6 logic low (0):
>echo 0 > /sys/class/gpio/gpio38/value
```

Accessing the GPIO files in C with Linux 3.2

To access the general purpose input/output pins and features from C, we use a similar technique to the Linux command technique. We access the pins and features via the file system described above via C file commands.

The C programming language has several helpful functions to open and establish a C file handle (fopen), set the file position indicator (fseek), write to a file (fprintf), read from a file (fscanf), insure data has been written to a file (fflush) and close a file (fclose) [Kelley and Pohl]. The provided file function descriptions below are admittedly brief. We only provide the information necessary to configure and use BeagleBone digital input and output pins. Descriptions were adapted from Kelley and Pohl's seminal book on the C programming language "A Book on C – Programming in C [Kelley and Pohl]."

- **fopen:** This function opens a file for use and attaches the file to a pointer. The file may be opened for reading ("r") or writing ("w"). If a file is used for reading, an infile pointer (ifp) should be attached to it. If used for writing an outfile (ofp) should be attached.

- **fseek:** This function is used with BeagleBone to set the file pointer to the beginning of the file.

- **fprintf:** This function writes variables of a specified type to a file. A wide variety of types may be written to the file: integers("%d"), characters ("%c"), strings ("%s"), and floating point ("%f").

- **fscanf:** This function reads variables of a specified type from a file. A wide variety of types may be read from the file: integers("%d"), characters ("%c"), strings ("%s"), and floating point ("%f").

- **fflush:** This function is used following a write operation (fprintf) to insure unwritten data has been written to a file.

- **fclose:** This function is used to close the specified file.

Several examples are provided to illustrate how these functions are used to configure Beagle-Bone's general purpose input and output (gpio) pins.

Example: In this example we redo the previous example of illuminating an LED from the Linux command line. An LED and a series connected resistor is connected to expansion header P8, pin3 (GPIO1_6 designated as gpio38).

```
//********************************************************************
//led1.cpp: illuminates an LED connected to expansion header P8, pin3
//(GPIO1_6 designated as gpio38)
//********************************************************************

#include <stdio.h>
#include <stddef.h>
#include <time.h>

#define  output "out"
#define  input  "in"

int  main (void)
{
//define file handles
FILE *ofp_gpmc_ad6, *ofp_export, *ofp_gpio38_value, *ofp_gpio38_direction;

//define pin variables
int mux_mode = 0x007, pin_number = 38, logic_status = 1;
char* pin_direction = output;

//configure the mux mode
ofp_gpmc_ad6 = fopen("/sys/kernel/debug/omap_mux/gpmc_ad6", "w");
if(ofp_gpmc_ad6 == NULL) {printf("Unable to open gpmc_ad6.\n");}
fseek(ofp_gpmc_ad6, 0, SEEK_SET);
fprintf(ofp_gpmc_ad6, "0x%02x", mux_mode);
fflush(ofp_gpmc_ad6);

//establish a direction and value file within export for gpio38
ofp_export = fopen("/sys/class/gpio/export", "w");
```

```
if(ofp_export == NULL) {printf("Unable to open export.\n");}
fseek(ofp_export, 0, SEEK_SET);
fprintf(ofp_export, "%d", pin_number);
fflush(ofp_export);

//configure gpio38 for writing
ofp_gpio38_direction = fopen("/sys/class/gpio/gpio38/direction", "w");
if(ofp_gpio38_direction==NULL){printf("Unable to open gpio38_direction.\n");}
fseek(ofp_gpio38_direction, 0, SEEK_SET);
fprintf(ofp_gpio38_direction, "%s",  pin_direction);
fflush(ofp_gpio38_direction);

//write a logic 1 to gpio38 to illuminate the LED
ofp_gpio38_value = fopen("/sys/class/gpio/gpio38/value", "w");
if(ofp_gpio38_value == NULL) {printf("Unable to open gpio38_value.\n");}
fseek(ofp_gpio38_value, 0, SEEK_SET);
fprintf(ofp_gpio38_value, "%d", logic_status);
fflush(ofp_gpio38_value);

//close all files
fclose(ofp_gpmc_ad6);
fclose(ofp_export);
fclose(ofp_gpio38_direction);
fclose(ofp_gpio38_value);
return 1;
}
//************************************************************************
```

Example: In this example we read the logic value of a pushbutton switch connected to expansion header P8, pin 5 (GPIO1_2 designated as gpio34). If the switch is logic high (1) an LED connected to expansion header P8, pin3 (GPIO1_6 designated as gpio38) is illuminated. A circuit diagram is provided in Figure 5.13. For BeagleBone black connect the LED to header P8 pin 13.

```
//************************************************************************
//led2.cpp: This program reads the logic value of a pushbutton switch
//connected to expansion header P8, pin 5 (GPIO1_2 designated as gpio34).
//If the switch is logic high (1) an LED connected to expansion header
//P8, pin3 (GPIO1_6 designated as gpio38) is illuminated.
//************************************************************************

#include <stdio.h>
```

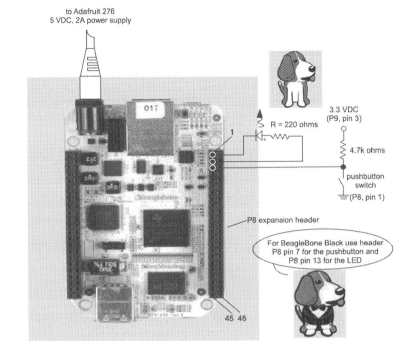

Figure 5.13: Interfacing an LED and switch to BeagleBone. **Note:** For BeagleBone black connect the LED to header P8 pin 13.

```
#include <stddef.h>
#include <time.h>

#define  output "out"
#define  input  "in"

int  main (void)
{
//define file handles for gpio38 (P8, pin 3, GPIO1_6)
FILE *ofp_gpmc_ad6, *ofp_export_38, *ofp_gpio38_value, *ofp_gpio38_direction;

//define file handles for gpio34 (P8, pin 5, GPIO1_2)
FILE *ofp_gpmc_ad2, *ofp_export_34, *ifp_gpio34_value, *ofp_gpio34_direction;

//define pin variables for gpio38
int mux_mode_38 = 0x0007, pin_number_38 = 38, logic_status_38 = 1;
```

```
char* pin_direction_38 = output;

//define pin variables for gpio34
int mux_mode_34 = 0x003f, pin_number_34 = 34, logic_status_34;
char* pin_direction_34 = input;

//gpio38 mux setting
ofp_gpmc_ad6 = fopen("/sys/kernel/debug/omap_mux/gpmc_ad6", "w");
if(ofp_gpmc_ad6 == NULL) {printf("Unable to open gpmc_ad6.\n");}
fseek(ofp_gpmc_ad6, 0, SEEK_SET);
fprintf(ofp_gpmc_ad6, "0x%02x", mux_mode_38);
fflush(ofp_gpmc_ad6);

//gpio34 mux setting
ofp_gpmc_ad2 = fopen("/sys/kernel/debug/omap_mux/gpmc_ad2", "w");
if(ofp_gpmc_ad2 == NULL) {printf("Unable to open gpmc_ad2.\n");}
fseek(ofp_gpmc_ad2, 0, SEEK_SET);
fprintf(ofp_gpmc_ad2, "0x%02x", mux_mode_34);
fflush(ofp_gpmc_ad2);

//create direction and value file for gpio38
ofp_export_38 = fopen("/sys/class/gpio/export", "w");
if(ofp_export_38 == NULL) {printf("Unable to open export.\n");}
fseek(ofp_export_38, 0, SEEK_SET);
fprintf(ofp_export_38, "%d", pin_number_38);
fflush(ofp_export_38);

//create direction and value file for gpio34
ofp_export_34 = fopen("/sys/class/gpio/export", "w");
if(ofp_export_34 == NULL) {printf("Unable to open export.\n");}
fseek(ofp_export_34, 0, SEEK_SET);
fprintf(ofp_export_34, "%d", pin_number_34);
fflush(ofp_export_34);

//configure gpio38 direction
ofp_gpio38_direction = fopen("/sys/class/gpio/gpio38/direction", "w");
if(ofp_gpio38_direction==NULL){printf("Unable to open gpio38_direction.\n");}
fseek(ofp_gpio38_direction, 0, SEEK_SET);
fprintf(ofp_gpio38_direction, "%s",  pin_direction_38);
```

```
fflush(ofp_gpio38_direction);

//configure gpio34 direction
ofp_gpio34_direction = fopen("/sys/class/gpio/gpio34/direction", "w");
if(ofp_gpio34_direction==NULL){printf("Unable to open gpio34_direction.\n");}
fseek(ofp_gpio34_direction, 0, SEEK_SET);
fprintf(ofp_gpio34_direction, "%s",  pin_direction_34);
fflush(ofp_gpio34_direction);

//configure gpio38 value \--- initially set logic high
ofp_gpio38_value = fopen("/sys/class/gpio/gpio38/value", "w");
if(ofp_gpio38_value == NULL) {printf("Unable to open gpio38_value.\n");}
fseek(ofp_gpio38_value, 0, SEEK_SET);
fprintf(ofp_gpio38_value, "%d", logic_status_38);
fflush(ofp_gpio38_value);

while(1)
  {
  //configure gpio34 value and read the gpio34 pin
  ifp_gpio34_value = fopen("/sys/class/gpio/gpio34/value", "r");
  if(ifp_gpio34_value == NULL) {printf("Unable to open gpio34_value.\n");}
  fseek(ifp_gpio34_value, 0, SEEK_SET);
  fscanf(ifp_gpio34_value, "%d", &logic_status_34);
  fclose(ifp_gpio34_value);
  printf("%d", logic_status_34);
  if(logic_status_34 == 1)
    {
    //set gpio38 logic high
    fseek(ofp_gpio38_value, 0, SEEK_SET);
    logic_status_38 = 1;
    fprintf(ofp_gpio38_value, "%d", logic_status_38);
    fflush(ofp_gpio38_value);
    printf("High\n");
    }
  else
    {
    //set gpio38 logic low
    fseek(ofp_gpio38_value, 0, SEEK_SET);
    logic_status_38 = 0;
```

```
    fprintf(ofp_gpio38_value, "%d", logic_status_38);
    fflush(ofp_gpio38_value);
    printf("Low\n");
    }
  }

//close files
fclose(ofp_gpmc_ad6);
fclose(ofp_export_38);
fclose(ofp_gpio38_direction);
fclose(ofp_gpio38_value);

fclose(ofp_gpmc_ad2);
fclose(ofp_export_34);
fclose(ofp_gpio34_direction);
fclose(ifp_gpio34_value);

return 1;
}
//***********************************************************************
```

Example: In this example we toggle an LED connected to expansion header P8, pin 3 (GPIO1_6 designated as gpio38) at five second intervals.

The program uses the "difftime" function. The definition for this function is included in the "time.h" header file. The "difftime" function prototype is [Kelley and Pohl]:

```
double difftime(time_t t0, time_t t1);
```

The "difftime" function computes the amount of elapsed time (t1–t0) in seconds and returns it as a double. A time hack may be obtained using:

```
//define time variable
time_t   now;

//get a time hack
now = time(NULL);
```

```
//***********************************************************************
//led3.cpp: this programs toggles (flashes) an LED connected to
//expansion header P8, pin 3 (GPIO1_6 designated as gpio38) at five
//second intervals.
//***********************************************************************
```

```c
#include <stdio.h>
#include <stddef.h>
#include <time.h>

#define  output "out"
#define  input  "in"

int  main (void)
{
//define file handles
FILE *ofp_gpm6_ad6, *ofp_export, *ofp_gpio38_value, *ofp_gpio38_direction;

//define pin variables
int mux_mode = 0x007, pin_number = 38, logic_status = 1;
char* pin_direction = output;

//time parameters
time_t  now, later;

ofp_gpm6_ad6 = fopen("/sys/kernel/debug/omap_mux/gpmc_ad6", "w");
if(ofp_gpm6_ad6 == NULL) {printf("Unable to open gpmc_ad6.\n");}
fseek(ofp_gpm6_ad6, 0, SEEK_SET);
fprintf(ofp_gpm6_ad6, "0x%02x", mux_mode);
fflush(ofp_gpm6_ad6);

ofp_export = fopen("/sys/class/gpio/export", "w");
if(ofp_export == NULL) {printf("Unable to open export.\n");}
fseek(ofp_export, 0, SEEK_SET);
fprintf(ofp_export, "%d", pin_number);
fflush(ofp_export);

ofp_gpio38_direction = fopen("/sys/class/gpio/gpio38/direction", "w");
if(ofp_gpio38_direction==NULL){printf("Unable to open gpio38_direction.\n");}
fseek(ofp_gpio38_direction, 0, SEEK_SET);
fprintf(ofp_gpio38_direction, "%s",  pin_direction);
fflush(ofp_gpio38_direction);

ofp_gpio38_value = fopen("/sys/class/gpio/gpio38/value", "w");
```

```
if(ofp_gpio38_value == NULL) {printf("Unable to open gpio38_value.\n");}
fseek(ofp_gpio38_value, 0, SEEK_SET);
logic_status = 1;
fprintf(ofp_gpio38_value, "%d", logic_status);
fflush(ofp_gpio38_value);

now = time(NULL);
later = time(NULL);

while(1)
  {
  while(difftime(later, now) < 5.0)
    {
    later = time(NULL); //keep checking time
    }
  if(logic_status == 1) logic_status = 0;
    else logic_status = 1;
  //write to gpio38
  fprintf(ofp_gpio38_value, "%d", logic_status);
  fflush(ofp_gpio38_value);
  now=time(NULL);
  later=time(NULL);
  }
fclose(ofp_gpm6_ad6);
fclose(ofp_export);
fclose(ofp_gpio38_direction);
fclose(ofp_gpio38_value);
return 1;
}


//***********************************************************************
```

In this section we have learned how to configure and use BeagleBone's general purpose input/output (gpio) features. The pins must be systematically configured for proper use. The status of configured pins may be monitored via the "/sys/kernel/debug/gpio" file. This file may be examined using the following Linux commands:

```
>cd /sys/kernel/debug
>less gpio
```

A helpful troubleshooting tool is the "strace" utility. It traces system calls and signals. For example, to install strace and use it to track the status of "led2.cpp", the following command sequence may be used:

```
>opkg install strace
>strace ./led2
```

Also, files may be transferred from BeagleBone to a USB drive inserted into the USB Beagle-Bone port. The USB drive must first be mounted to the BeagleBone Linux file structure and then the file may be transferred. The following command sequence accomplishes this task:

```
>mount /dev/sda1 /media/sda1
>cp <filename.cpp> /media/sda1/<filename.cpp>
```

In the next section we investigate how to perform an analog–to–digital conversion from the Linux command line and also from within a C program.

5.6   EXPANSION INTERFACE BEAGLEBONE BLACK

In this section we describe the expansion interface for BeagleBone Black and the device tree overlay system used to access the header pins. The expansion interface is illustrated in Figure 5.14. Also illustrated is cape header pins [www.beagleboard.org].

5.6.1   ACCESSING PINS WITH DEVICE TREE OVERLAYS –LINUX 3.8

In this section we provide a brief introduction to the device tree overlay concept employed with BeagleBone Black equipped a 3.8 Linux kernel from the most recent release of the Angstrom Distribution. We discuss the motivation for the device tree overlay approach, some related general concepts, the basic format of a device tree overlay, review the overlay for BeagleBone black, and conclude with an example. Information for this section was provided by a series of articles listed in the References section. The reader is encouraged to review these articles in detail [Devicetree, Likely, Hipster].

5.6.2   OVERVIEW

The device tree is a software data structure used to describe the hardware configuration of a specific processor to Linux. It allows the Linux kernel to remain the same even when used on a wide variety of processor configurations. Hardware specific details are passed to the Linux operating system via a device tree written specifically for the hardware device. The device tree is read via Linux during the boot operation. It may also be modified during run time by an application executing on the processor.

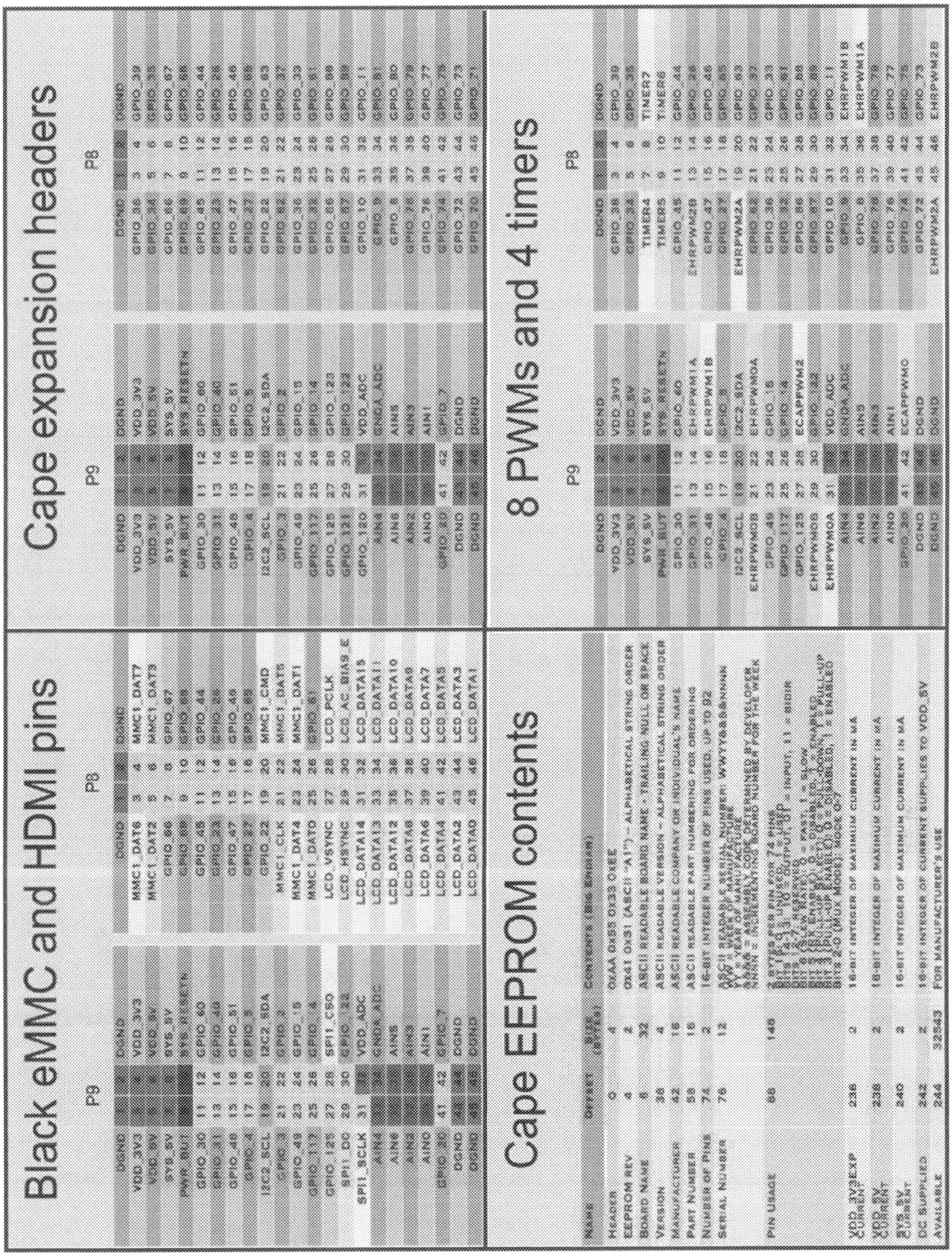

Figure 5.14: BeagleBone Black expansion interface [www.beagleboard.org].

5.6.3   BINARY TREE

The device tree is a software data structure similar to a binary tree. A binary tree consists of a collection of nodes as shown in Figure 5.15 [Korsch and Garrett]. Each node may hold a variety of data and also two links to its successor or child nodes. For example, Node 3 is the child or successor of Node 1. While Nodes 6 and 7 are child nodes of Node 3. From Node 3's point of view, Node 1 is its predecessor or parent node.

Access is gained to the binary tree via its root node. The tree may then be traversed to obtain the information at each node as required.

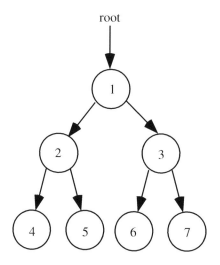

Figure 5.15: Binary tree (adapted from Korsch and Garrett).

5.6.4   DEVICE TREE FORMAT

In this section we provide a brief introduction to the device tree format. An excellent, detailed discussion of this topic is provided at www.devicetree.org. The device tree uses a binary tree format to contain device hardware properties at each node along with connections to its child nodes. The device tree root is designated with a forward slash (/) symbol. It then has a compatibility statement to indicate the hardware processor in use. Following the compatibility statement is a listing of nodes. Each node contains information about specific hardware devices aboard the processor. The tree hierarchy also represents how the devices are interconnected within the processor.

Each of the individual nodes contains a compatibility statement and device addressing parameters. The device addressing parameters include the base address for its associated registers and the number of registers used. Provided in the code snapshot below is the basic format of the device tree [devicetree].

```
//******************************************************************
/{
 compatible = ''<manufacturer>,<model>";


 <node name>[@<device address>]{
     compatible = ''<manufacturer>,<model>";
     reg = <start address   length>;
     };

//Additional device node descriptions


 :
 :
 :


 };

//******************************************************************
```

The device tree is loaded during system boot. However, portions of the device tree may be modified by using a fragment modifier during the boot process or during program execution.

5.6.5   BEAGLEBONE DEVICE TREE–LINUX 3.8

Pantellis has provided a device tree overlay for BeagleBone users. Pantelis also introduced a BeagleBone cape manager (capemgr) to assist with updating BeagleBone device tree information boot time or during program run time [Pantellis].

Example: The following example illustrates how the device tree may be modified using a fragment statement. We start by defining the two key sysfs entries we'll use to check status. An interface to the capemgr is provided by 'slots.' It notifies capemgr to load additional device tree overlay fragments and also reports on what has already been loaded. It uses the EEPROMs on the cape plug-in boards to identify the board [Kridner].

```
export SLOTS=/sys/devices/bone_capemgr.8/slots
export PINS=/sys/kernel/debug/pinctrl/44e10800.pinmux/pins
```

Provided below is a device tree overlay to update the multiplexer setting of header P9 pin 42 to mode 7.

```
//******************************************************************
//devicetree overlay (pinmux-test-7.dts):
```

```
//
//Copyright (C) 2012 Texas Instruments Incorporated - http://www.ti.com/
//
//This program is free software; you can redistribute it and/or modify
//it under the terms of the GNU General Public License version 2 as
//published by the Free Software Foundation.
//*********************************************************************

/dts-v1/;
/plugin/;

/ {
    compatible = "ti,beaglebone", "ti,beaglebone-black";

    /* identification */
    part-number = "pinctrl-test-7";

    fragment@0 {
        target = <&am33xx_pinmux>;
        __overlay__ {
            pinctrl_test: pinctrl_test_7_pins {
                pinctrl-single,pins = <
                    0x164 0x07  /* P9_42 muxRegOffset, OUTPUT | MODE7 */
                >;
            };
        };
    };

    fragment@1 {
        target = <&ocp>;
        __overlay__ {
            test_helper: helper {
                compatible = "bone-pinmux-helper";
                pinctrl-names = "default";
                pinctrl-0 = <&pinctrl_test>;
                status = "okay";
            };
        };
    };
```

```
};
//*********************************************************************
```

The device tree fragments are now compiled and installed in /lib/firmware. In this example it is performed natively on the BeagleBone.

```
dtc -O dtb -o pinctrl-test-7-00A0.dtbo -b 0 -@ pinctrl-test-7.dts
dtc -O dtb -o pinctrl-test-0-00A0.dtbo -b 0 -@ pinctrl-test-0.dts
cp pinctrl-test-7-00A0.dtbo /lib/firmware/
cp pinctrl-test-0-00A0.dtbo /lib/firmware/
```

Before continuing with the example, let's check out starting point state.

```
# cat $SLOTS
 0: 54:PF---
 1: 55:PF---
 2: 56:PF---
 3: 57:PF---
 4: ff:P-O-L Bone-LT-eMMC-2G,00A0,Texas Instrument,BB-BONE-EMMC-2G
 5: ff:P-O-L Bone-Black-HDMI,00A0,Texas Instrument,BB-BONELT-HDMI
```

As can be seen from the report, slots 0–3 get assigned by EEPROM IDs on the capes. There are 4 possible addresses for the EEPROMs (typically determined by switches on the boards) enabling up to 4 boards to be stacked, depending on what functions they use. Additional slots are "virtual," added incrementally and are triggered in other ways. In the above report you can see that no capes are currently installed. There are two "virtual" capes installed, one for the HDMI and one for the eMMC. It makes sense to manage these as capes because both interfaces consume pins on the cape bus. These two "virtual" capes are triggered on BeagleBone Black that includes the eMMC and HDMI on the board. Disabling these capes would enable other capes to make use of their pins.

In the next step, let's tell the capemgr to load our device tree overlay fragment that configures the target pin's pinmux. Carefully take a look at the messages that are produced by the kernel and review the capemgr status and the status of the pinmux.

```
# echo pinctrl-test-7 > $SLOTS
# dmesg | tail
[   65.323606] bone-capemgr bone_capemgr.8:
    part_number 'pinctrl-test-7', version 'N/A'
[   65.323744] bone-capemgr bone_capemgr.8:
    slot #6: generic override
[   65.323794] bone-capemgr bone_capemgr.8:
    bone: Using override eeprom data at slot 6
[   65.323845] bone-capemgr bone_capemgr.8:
    slot #6: 'Override Board Name,00A0,Override Manuf,pinctrl-test-7'
```

```
[   65.324201] bone-capemgr bone_capemgr.8:
    slot #6: Requesting part number/version based
    'pinctrl-test-7-00A0.dtbo
[   65.325712] bone-capemgr bone_capemgr.8:
    slot #6: Requesting firmware 'pinctrl-test-7-00A0.dtbo'
    for board-name 'Override Board Name', version '00A0'
[   65.326239] bone-capemgr bone_capemgr.8:
    slot #6: dtbo 'pinctrl-test-7-00A0.dtbo' loaded;
    converting to live tree
[   65.327973] bone-capemgr bone_capemgr.8: slot #6: #2 overlays
[   65.338533] bone-capemgr bone_capemgr.8: slot #6: Applied #2 overlays.
# cat $SLOTS
 0: 54:PF---
 1: 55:PF---
 2: 56:PF---
 3: 57:PF---
 4: ff:P-O-L Bone-LT-eMMC-2G,00A0,Texas Instrument,BB-BONE-EMMC-2G
 5: ff:P-O-L Bone-Black-HDMI,00A0,Texas Instrument,BB-BONELT-HDMI
 6: ff:P-O-L Override Board Name,00A0,Override Manuf,pinctrl-test-7
```

The $PINS file is a debug entry for the pinctrl kernel module. It provides the status of the pinmux. To examine a specific pin state, grep for the lower bits of the address where the pinmux control register is located. The Linux grep command searches the given file for lines containing a match. As you can see, the mux mode is now 7.

```
# cat $PINS | grep 964
pin 89 (44e10964) 00000007 pinctrl-single
```

Now we"ll have the capemgr unload the overlay so that a different one can be loaded.

```
# A=`perl -pe 's/^.*(\d+):.*/$1/' $SLOTS | tail -1`
# echo "-$A"
-6
# echo "-$A" > $SLOTS
# dmesg | tail
[   73.517002] bone-capemgr bone_capemgr.8: Removed slot #6
[   73.517002] bone-capemgr bone_capemgr.8: Removed slot #6
And then tell capemgr to load an alternative overlay.
# echo pinctrl-test-0 > $SLOTS
# dmesg | tail
[   73.663144] bone-capemgr bone_capemgr.8:
    part_number 'pinctrl-test-0', version 'N/A'
```

```
[   73.663207] bone-capemgr bone_capemgr.8:
    slot #7: generic override
[   73.663226] bone-capemgr bone_capemgr.8:
    bone: Using override eeprom data at slot 7
[   73.663244] bone-capemgr bone_capemgr.8:
    slot #7: 'Override Board Name,00A0,Override Manuf,pinctrl-test-0'
[   73.663340] bone-capemgr bone_capemgr.8:
    slot #7: Requesting part number/version based
    'pinctrl-test-0-00A0.dtbo
[   73.663357] bone-capemgr bone_capemgr.8:
    slot #7: Requesting firmware 'pinctrl-test-0-00A0.dtbo'
    for board-name 'Override Board Name', version '00A0'
[   73.663602] bone-capemgr bone_capemgr.8: slot #7:
    dtbo 'pinctrl-test-0-00A0.dtbo' loaded; converting to live tree
[   73.663857] bone-capemgr bone_capemgr.8:
    slot #7: #2 overlays
[   73.674682] bone-capemgr bone_capemgr.8:
    slot #7: Applied #2 overlays.
```

And what does the pinmux look like now?

```
# cat $PINS | grep 964
pin 89 (44e10964) 00000000 pinctrl-single
```

Fortunately, there are existing devicetree fragments you can load to avoid creating these fragments yourself. Further, the default configurations are useful in many cases. The device tree for BeagleBone is provided in Appendix D.

5.7   FUNDAMENTAL EXAMPLES PROGRAMMING IN C WITH BEAGLEBONE BLACK–LINUX 3.8

In this section we repeat the basic examples provided earlier in the chapter for the original Beagle-Bone. They have been adapted for execution on BeagleBone Black with Linux–3.8.

```
//*****************************************************************
//led1.cpp: illuminates an LED connected to expansion header P8, pin13
//(GPIO1_12 designated as gpio44)
//*****************************************************************

#include <stdio.h>
#include <stddef.h>
#include <time.h>
```

```c
#define  output "out"
#define  input  "in"

int  main (void)
{
//define file handles
FILE *ofp_export, *ofp_gpio44_value, *ofp_gpio44_direction;

//define pin variables
int pin_number = 44, logic_status = 0;
char* pin_direction = output;

//establish a direction and value file within export for gpio44
ofp_export = fopen("/sys/class/gpio/export", "w");
if(ofp_export == NULL) {printf("Unable to open export.\n");}
fseek(ofp_export, 0, SEEK_SET);
fprintf(ofp_export, "%d", pin_number);
fflush(ofp_export);

//configure gpio44 for writing
ofp_gpio44_direction = fopen("/sys/class/gpio/gpio44/direction", "w");
if(ofp_gpio44_direction==NULL){printf("Unable to open gpio44_direction.\n");}
fseek(ofp_gpio44_direction, 0, SEEK_SET);
fprintf(ofp_gpio44_direction, "%s",  pin_direction);
fflush(ofp_gpio44_direction);

//write a logic 1 to gpio44 to illuminate the LED
ofp_gpio44_value = fopen("/sys/class/gpio/gpio44/value", "w");
if(ofp_gpio44_value == NULL) {printf("Unable to open gpio44_value.\n");}
fseek(ofp_gpio44_value, 0, SEEK_SET);
fprintf(ofp_gpio44_value, "%d", logic_status);
fflush(ofp_gpio44_value);

//close all files
fclose(ofp_export);
fclose(ofp_gpio44_direction);
fclose(ofp_gpio44_value);
return 1;
```

```cpp
}
//***********************************************************************


//***********************************************************************
//led2.cpp: This program reads the logic value of a pushbutton switch
//connected to expansion header P8, pin 7 (GPIO2_2 designated as gpio66).
//If the switch is logic high (1) an LED connected to expansion header
//P8, pin13 (GPIO1_12 designated as gpio44) is illuminated.
//***********************************************************************

#include <stdio.h>
#include <stddef.h>
#include <time.h>

#define  output "out"
#define  input  "in"

int  main (void)
{
//define file handles for gpio44 (P8, pin 13, GPIO1_12)
FILE *ofp_export_44, *ofp_gpio44_value, *ofp_gpio44_direction;

//define file handles for gpio66 (P8, pin 7, GPIO2_2)
FILE *ofp_export_66, *ifp_gpio66_value, *ofp_gpio66_direction;

//define pin variables for gpio44
int pin_number_44 = 44, logic_status_44 = 1;
char* pin_direction_44 = output;

//define pin variables for gpio66
int pin_number_66 = 66, logic_status_66;
char* pin_direction_66 = input;

//create direction and value file for gpio44
ofp_export_44 = fopen("/sys/class/gpio/export", "w");
if(ofp_export_44 == NULL) {printf("Unable to open export.\n");}
fseek(ofp_export_44, 0, SEEK_SET);
fprintf(ofp_export_44, "%d", pin_number_44);
fflush(ofp_export_44);
```

```
//create direction and value file for gpio66
ofp_export_66 = fopen("/sys/class/gpio/export", "w");
if(ofp_export_66 == NULL) {printf("Unable to open export.\n");}
fseek(ofp_export_66, 0, SEEK_SET);
fprintf(ofp_export_66, "%d", pin_number_66);
fflush(ofp_export_66);

//configure gpio44 direction
ofp_gpio44_direction = fopen("/sys/class/gpio/gpio44/direction", "w");
if(ofp_gpio44_direction == NULL) {printf("Unable to open gpio44_direction.\n");
fseek(ofp_gpio44_direction, 0, SEEK_SET);
fprintf(ofp_gpio44_direction, "%s",  pin_direction_44);
fflush(ofp_gpio44_direction);

//configure gpio66 direction
ofp_gpio66_direction = fopen("/sys/class/gpio/gpio66/direction", "w");
if(ofp_gpio66_direction == NULL) {printf("Unable to open gpio66_direction.\n");
fseek(ofp_gpio66_direction, 0, SEEK_SET);
fprintf(ofp_gpio66_direction, "%s",  pin_direction_66);
fflush(ofp_gpio66_direction);

//configure gpio44 value \--- initially set logic high
ofp_gpio44_value = fopen("/sys/class/gpio/gpio44/value", "w");
if(ofp_gpio44_value == NULL) {printf("Unable to open gpio44_value.\n");}
fseek(ofp_gpio44_value, 0, SEEK_SET);
fprintf(ofp_gpio44_value, "%d", logic_status_44);
fflush(ofp_gpio44_value);

while(1)
  {
  //configure gpio66 value and read the gpio66 pin
  ifp_gpio66_value = fopen("/sys/class/gpio/gpio66/value", "r");
  if(ifp_gpio66_value == NULL) {printf("Unable to open gpio66_value.\n");}
  fseek(ifp_gpio66_value, 0, SEEK_SET);
  fscanf(ifp_gpio66_value, "%d", &logic_status_66);
  fclose(ifp_gpio66_value);
  printf("%d", logic_status_66);
  if(logic_status_66 == 1)
```

```
   {
   //set gpio44 logic high
   fseek(ofp_gpio44_value, 0, SEEK_SET);
   logic_status_44 = 1;
   fprintf(ofp_gpio44_value, "%d", logic_status_44);
   fflush(ofp_gpio44_value);
   printf("High\n");
   }
 else
   {
   //set gpio44 logic low
   fseek(ofp_gpio44_value, 0, SEEK_SET);
   logic_status_44 = 0;
   fprintf(ofp_gpio44_value, "%d", logic_status_44);
   fflush(ofp_gpio44_value);
   printf("Low\n");
   }
 }

//close files
fclose(ofp_export_44);
fclose(ofp_gpio44_direction);
fclose(ofp_gpio44_value);

fclose(ofp_export_66);
fclose(ofp_gpio66_direction);
fclose(ifp_gpio66_value);

return 1;
}
//*****************************************************************


//*****************************************************************
//led3.cpp: this programs toggles (flashes) an LED connected to
//expansion header P8, pin 13 (GPIO1_12 designated as gpio44) at five
//second intervals.
//*****************************************************************

#include <stdio.h>
```

```c
#include <stddef.h>
#include <time.h>

#define  output "out"
#define  input  "in"

int  main (void)
{
//define file handles
FILE *ofp_export, *ofp_gpio44_value, *ofp_gpio44_direction;

//define pin variables
int pin_number = 44, logic_status = 1;
char* pin_direction = output;

//time parameters
time_t  now, later;

ofp_export = fopen("/sys/class/gpio/export", "w");
if(ofp_export == NULL) {printf("Unable to open export.\n");}
fseek(ofp_export, 0, SEEK_SET);
fprintf(ofp_export, "%d", pin_number);
fflush(ofp_export);

ofp_gpio44_direction = fopen("/sys/class/gpio/gpio44/direction", "w");
if(ofp_gpio44_direction == NULL) {printf("Unable to open gpio44_direction.\n");
fseek(ofp_gpio44_direction, 0, SEEK_SET);
fprintf(ofp_gpio44_direction, "%s",  pin_direction);
fflush(ofp_gpio44_direction);

ofp_gpio44_value = fopen("/sys/class/gpio/gpio44/value", "w");
if(ofp_gpio44_value == NULL) {printf("Unable to open gpio44_value.\n");}
fseek(ofp_gpio44_value, 0, SEEK_SET);
logic_status = 1;
fprintf(ofp_gpio44_value, "%d", logic_status);
fflush(ofp_gpio44_value);

now = time(NULL);
later = time(NULL);
```

```
while(1)
  {
  while(difftime(later, now) < 5.0)
    {
    later = time(NULL); //keep checking time
    }
  if(logic_status == 1) logic_status = 0;
    else logic_status = 1;
  //write to gpio44
  fprintf(ofp_gpio44_value, "%d", logic_status);
  fflush(ofp_gpio44_value);
  now=time(NULL);
  later=time(NULL);
  }
fclose(ofp_export);
fclose(ofp_gpio44_direction);
fclose(ofp_gpio44_value);
return 1;
}


//*************************************************************************
```

For the remainder of the chapter we review exposed functions aboard the BeagleBone. For each function we provide a bit of theory and then provide examples for BeagleBone original operating Linux 3.2 and BeagleBone Black equipped with Linux 3.8.

5.8   ANALOG–TO–DIGITAL CONVERTERS (ADC)

A processor may be used to capture analog information from the natural world, determine a course of action based on the information collected and the resident algorithm, and issue control signals to implement the decision. Information from the natural world, is analog or continuous in nature; whereas, a processor is digital. A subsystem to convert an analog signal to a digital form is required. An ADC system performs this task while a digital–to–analog converter (DAC) performs the conversion in the opposite direction.

In this section we discuss the ADC conversion process followed by a discussion of the successive–approximation ADC technique used aboard BeagleBone. We then review the basic features of the BeagleBone ADC system. We conclude our ADC discussion with several illustrative code examples.

5.8.1   ADC PROCESS: SAMPLING, QUANTIZATION AND ENCODING

In this section, we provide an abbreviated discussion of the ADC process. This discussion was condensed from "Atmel AVR Microcontroller Primer Programming and Interfacing." The interested reader is referred to this text for additional details and examples [Barrett and Pack]. We begin with three important stages associated with the ADC: sampling, quantization, and encoding.

Sampling. Sampling is the process of taking 'snap shots' of a signal over time. When we sample a signal, we want to minimize the number of samples taken while retaining the capability to reconstruct the original signal from the samples. Intuitively, the rate of change in signal behavior determines the sample interval required to faithfully reconstruct the signal. The bottom line: we must use the appropriate sampling rate to capture the analog signal for a faithful representation in digital systems.

Harry Nyquist from Bell Laboratory studied the sampling process and derived a criterion that determines the minimum sampling rate for a continuous analog signal. The minimum sampling rate derived is known as the Nyquist sampling rate. It states a signal must be sampled at least twice as fast as the highest frequency content of the signal of interest. For example, the human voice signal contains frequency components that span from approximately 20 Hz to 4 kHz. The Nyquist sample theorem requires sampling the signal at least at 8 kHz, 8,000 'snap shots' every second. Also, when a signal is sampled, a low pass anti–aliasing filter must be used to insure the Nyquist sampling rate is not violated. In the human voice example, a low pass filter with a cutoff frequency of 4 kHz would be used before the sampling circuitry for this purpose.

Quantization. In many digital systems, the incoming signals are voltage signals. The voltage signals are obtained from physical variables (pressure, temperature, etc.) via transducers, such as microphones, angle sensors, and infrared sensors. The voltage signals are then conditioned to map their range with the input range of the digital system. In the case of BeagleBone, the analog signal must be conditioned such that it does not exceed 1.8 VDC. Techniques for signal conditioning were discussed earlier in the book.

When an analog signal is sampled, the digital system needs a means to represent the captured samples. The quantization of a sampled analog signal is represented as one of the digital quantization levels. For example, if you have three bits available for quantization, you can represent eight different levels: 000, 001, 010, 011, 100, 101, 110, and 111. In general, given n bits, we have 2^n unique numbers or quantization levels in our system. Figure 5.16 shows how n bits are used to quantize a range of values. As the number of bits used for the quantization levels increase for a given input range, the 'distance' between two adjacent levels decreases accordingly. Intuitively, the more quantization levels means the better mapping of an incoming signal to its true value.

Encoding. The encoding process converts a quantized signal into a digital binary number. Suppose again we are using eight bits to quantize a sampled analog signal. The quantization levels are determined by the eight bits and each sampled signal is quantized as one of 256 quantization levels. Consider the two sampled signals shown in Figure 5.16. The first sample is mapped to quantization level 2 and the second one is mapped to quantization level 198. Note the amount of quantization

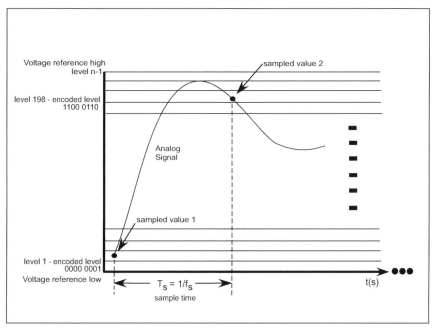

Figure 5.16: Sampling, quantization, and encoding [Barrett and Pack].

error introduced for both samples. The quantization error is inversely proportional to the number of bits used to quantize the signal.

5.8.2   RESOLUTION AND DATA RATE

Resolution. Resolution is a measure used to quantize an analog signal. In fact, resolution is nothing more than the voltage 'distance' between two adjacent quantization levels we discussed earlier. That is, as we increase the available number of quantization levels within a fixed voltage range, the distance between adjacent levels decreases, reducing the quantization error of a sampled signal. As the number of quantization levels increase, the error decreases, making the representation of a sampled analog signal more accurate in the corresponding digital form. The number of bits used for the quantization is directly proportional to the resolution of a system. In general, resolution may be defined as:

$$resolution = (voltage\ span)/2^b = (V_{ref\ high} - V_{ref\ low})/2^b$$

for BeagleBone, the resolution is:

$$resolution = (1.8 - 0)/2^{12} = 1.8/4096 = 439.45\ \mu V$$

Data rate. Data rate is the amount of data generated by a system per unit time. Typically, the number of bits or the number of bytes per second is used as the data rate of a system. In the previous section, we observed the more bits we use for the quantization levels, the more accurate we can represent a sampled analog signal. So why not use the maximum number of bits when we convert analog signals to digital counterparts? For example, suppose you are working for a telephone company and your switching system must accommodate 100,000 customers. For each individual phone conversation, suppose the company uses an 8 kHz sampling rate (f_s) and 10 bits for the quantization levels for each sampled signal.[1] This means the voice conversation will be sampled every 125 microseconds (T_s) due to the reciprocal relationship between (f_s) and (T_s). If all customers are making out of town calls, what is the number of bits your switching system must process to accommodate all calls? The answer will be 100,000 x 8,000 x 10 or eight billion bits every second! For these reasons, when making decisions on the number of bits used for the quantization levels and the sampling rate, you must consider the computational burden the selection will produce on the computational capabilities of a digital system versus the required system resolution.

Dynamic range. Dynamic range is a measure used to describe the signal to noise ratio. The unit used for measurement is the Decibel (dB), which is the strength of a signal with respect to a reference signal. The greater the dB number, the stronger the signal is compared to a noise signal. The definition of the dynamic range is $20 \, log \, 2^b$ where b is the number of bits used to convert analog signals to digital signals. Typically, you will find 8 to 12 bits used in commercial analog–to–digital converters, translating the dynamic range from $20 \, log \, 2^8$ dB to $20 \, log \, 2^{12}$ dB.

5.8.3   ADC CONVERSION TECHNOLOGIES

The ARM processor aboard BeagleBone uses a successive approximation converter technique to convert an analog sample into a 12–bit digital representation. The digital value is typically represented as an integer value between 0 and 4095. In this section, we discuss this type of conversion process. For a review of other converter techniques, the interested reader is referred to "Atmel AVR Microcontroller Primer: Programming and Interfacing." In certain applications, you are required to use converters external to the processor.

The successive–approximation technique uses a digital–to–analog converter, a controller, and a comparator to perform the ADC process. Starting from the most significant bit down to the least significant bit, the controller turns on each bit one at a time and generates an analog signal, with the help of the digital–to–analog converter, to be compared with the original input analog signal. Based on the result of the comparison, the controller changes or leaves the current bit and turns on the next most significant bit. The process continues until decisions are made for all available bits. Figure 5.17 shows the architecture of this type of converter. The advantage of this technique is that the conversion time is uniform for any input, but the disadvantage of the technology is the use of complex hardware for implementation.

---

[1] We ignore overhead involved in processing a phone call such as multiplexing, de–multiplexing, and serial–to–parallel conversion.

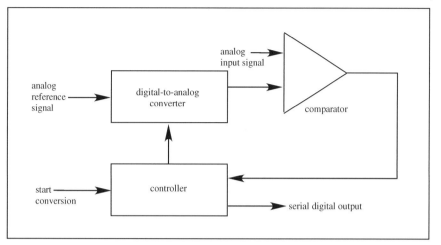

Figure 5.17: Successive-approximation ADC [Barrett and Pack].

5.8.4   BEAGLEBONE ADC SUBSYSTEM DESCRIPTION –LINUX 3.2

BeagleBone is equipped with an eight channel ADC system. The maximum input voltage for each of the channels is 1.8 VDC. The onboard ADC system uses the successive approximation conversion technique to convert an analog sample into a 12–bit digital value.

Access to the eight analog channels is provided via the following expansion header P9 pins:

- AIN0 pin 39
- AIN1 pin 40
- AIN2 pin 37
- AIN3 pin 38
- AIN4 pin 33
- AIN5 pin 36
- AIN6 pin 35
- Analog ground, GNDA_ADC, pin 34

As with the general–purpose digital input/output (GPIO) pins, access to the analog pin values are via dedicated files. The following files provide access to each of the analog input pins:

```
AIN0 pin 39, file: /sys/bus/platform/devices/tsc/ain1
AIN1 pin 40, file: /sys/bus/platform/devices/tsc/ain2
```

```
AIN2 pin 37, file: /sys/bus/platform/devices/tsc/ain3
AIN3 pin 38, file: /sys/bus/platform/devices/tsc/ain4
AIN4 pin 33, file: /sys/bus/platform/devices/tsc/ain5
AIN5 pin 36, file: /sys/bus/platform/devices/tsc/ain6
AIN6 pin 35, file: /sys/bus/platform/devices/tsc/ain7
```

5.8.5   ADC CONVERSION VIA LINUX 3.2

The ADC inputs may be accessed via the Linux command line. Provided in Figure 5.18a) is an analog input test configuration. A trimmer potentiometer is set for 1.5 VDC. The trimmer's wiper connection may be connected to ADC input AIN0 (P9, pin 39). To access the AIN0 analog value the following command may be used:

```
>cat /sys/bus/platform/devices/tsc/ain1
>3399
```

   Note: To access pin AIN0 the ain1 file is used. A value of 3399 is reported. What does this mean? The ADC system aboard BeagleBone is 12 bits with a maximum conversion value of 1.8 VDC. Therefore, a reading of 3399 equates to:

$$3399/(2^{12}) \times 1.8 = 1.4937 \; volts$$

In Figure 5.18b) a test configuration is provided to characterize the response of a Sharp GP2D12 IR sensor used on the Blinky 602A robot. The command provided above may be used repeatedly to characterize the sensor at different ranges. Rather than retype the command for each separate reading, the [Ctrl–P] followed by [Enter] automatically repeats and executes the command. Results of characterizing the IR sensor are provided in Figure 5.19.

   Provided in Figure 5.18c) is a test configuration for an LM34 precision Fahrenheit temperature sensor. The LM34 provides an output voltage that is linearly related to Fahrenheit temperature. The LM34 is rated over a -50° to $+300^o$F range. The LM34 provides an output of $+10.0 \; mV/^oF$.

5.8.6   ADC SUPPORT FUNCTIONS IN C LINUX 3.2

The ADC features may be accessed from within a C program using techniques similar to those provided for the GPIO pins. The next example illustrates how to access an ADC related pin via the file system.

   Example: In this example we read the analog value on AIN0 (P9, pin 39) provided by the LM34 temperature sensor. The connection for the LM34 to BeagleBone is provided in Figure 5.18c). The file used to read AIN0 (ain1) is first opened (fopen) and then read (fscanf). The value from the LM34 is printed to the terminal (printf). The temperature may be increased by holding the LM34 between your fingers or decreased using compressed air. The readings are accomplished continuously until [Control][C] is used to stop the program.

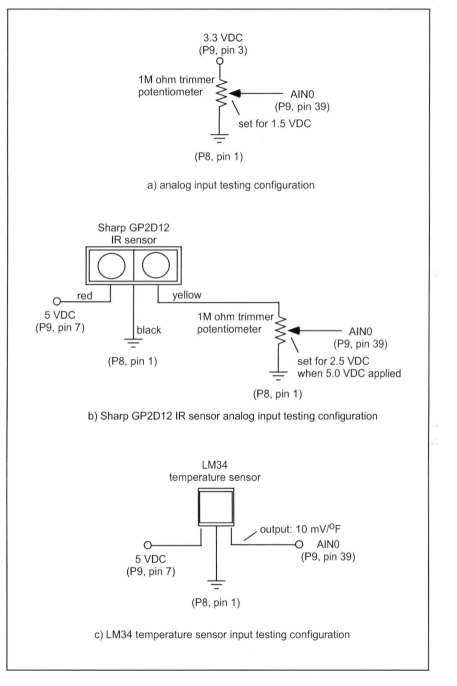

Figure 5.18: ADC test configurations.

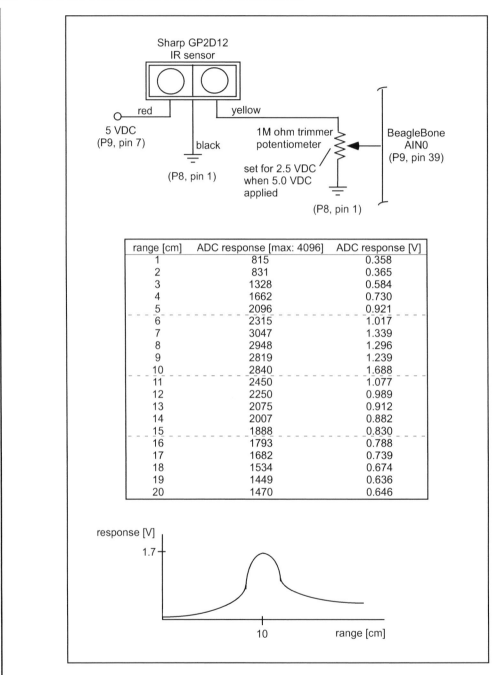

Figure 5.19: IR sensor characterization.

```cpp
//****************************************************************
//adc1.cpp: the analog value on AIN0 (P9, pin 39) provided by
//a LM34 temperature sensor is read continuously until
//[Control][C] is used to stop the program.
//****************************************************************

#include <stdio.h>
#include <stddef.h>
#include <time.h>
#include <math.h>

#define  output "out"
#define  input  "in"

int  main (void)
{
//define file handles
FILE *ifp_ain1;
float ain0_value;

ifp_ain1 = fopen("/sys/bus/platform/devices/tsc/ain1", "r");
if (ifp_ain1 == NULL) {printf("Unable to ain1.\n");}

while(1)
  {
  fseek(ifp_ain1, 0, SEEK_SET);
  fscanf(ifp_ain1, "%f", &ain0_value);
  printf("%f\n", ain0_value);
  }

fclose(ifp_ain1);
return 1;
}
//****************************************************************
```

Example: In this example readings of the LM34 are taken at two second intervals. The value from AIN0 (ain1) is read and printed to the terminal. Each reading is converted to a voltage and also a temperature. A #define statement is used to link the file path for ain1 to a more convenient (and shorter) name. The #define statements could be moved to a header file.

```cpp
//****************************************************************
```

```c
//include files
#include <stdio.h>
#include <stddef.h>
#include <time.h>
#include <math.h>

//define
#define  output "out"
#define  input  "in"
#define  ain1_in "/sys/bus/platform/devices/tsc/ain1"

//function prototypes
void delay_sec(float delay_value);

int  main (void)
{
//define file handles
FILE *ifp_ain1;
float ain0_value;
float ain0_voltage;
float ain0_temp;

ifp_ain1 = fopen(ain1_in, "r");
if (ifp_ain1 == NULL) {printf("Unable to ain1.\n");}

while(1)
  {
  fseek(ifp_ain1, 0, SEEK_SET);
  fscanf(ifp_ain1, "%f", &ain0_value);
  printf("AIN0 reading [of 4096]: %f\n", ain0_value);
  ain0_voltage = ((ain0_value/4096.0) * 1.8);
  printf("AIN0 voltage [V]: %f\n", ain0_voltage);
  ain0_temp = (ain0_voltage/.010);
  printf("AIN0 temperature [F]: %f\n\n", ain0_temp);
  delay_sec(2.0);
  }
fclose(ifp_ain1);
return 1;
}
```

```
//****************************************************************
//function definitions
//****************************************************************

void delay_sec(float delay_value)
{
time_t  now, later;

now = time(NULL);
later = time (NULL);

while(difftime(later, now) < delay_value)
  {
  later = time(NULL);                //keep checking time
  }
}


//****************************************************************
```

5.8.7   ADC SUPPORT FUNCTIONS IN C LINUX 3.8

Provided below are two examples using the ADC system with BeagleBone Black in Linux–3.8.

```
//****************************************************************
//adc1.cpp: the analog value on AIN0 (P9, pin 39) provided by
//a LM34 temperature sensor is read continuously until
//[Control][C] is used to stop the program.
//
//Before the AIN values show up, you must perform:
//  echo cape-bone-iio > /sys/devices/bone_capemgr.*/slots
//
//Notes:
// * The .''11" after helper can change between software revs
// * The scale of values is now 0-1800 (mV) instead of 0-4095
// * The AINx indexes now match the hardware documentation
//   instead of being off by 1
//****************************************************************

#include <stdio.h>
#include <stddef.h>
```

```
#include <time.h>
#include <math.h>

#define  output "out"
#define  input  "in"

int  main (void)
{
//define file handles
FILE *ifp_ain0;
float ain0_value;

ifp_ain0 = fopen("/sys/module/bone_iio_helper/drivers/
          platform:bone-iio-helper/helper.11/AIN0", "r");
if (ifp_ain0 == NULL) {printf("Unable to AIN0.\n");}

while(1)
  {
  fseek(ifp_ain0, 0, SEEK_SET);
  fscanf(ifp_ain0, "%f", &ain0_value);
  printf("%f\n", ain0_value);
  }

fclose(ifp_ain0);
return 1;
}
//***************************************************************

//***************************************************************
//include files
#include <stdio.h>
#include <stddef.h>
#include <time.h>
#include <math.h>

//define
#define  output "out"
#define  input  "in"
#define  ain0_in "/sys/module/bone_iio_helper/drivers/
               platform:bone-iio-helper/helper.11/AIN0"
```

```c
//function prototypes
void delay_sec(float delay_value);

int  main (void)
{
//define file handles
FILE *ifp_ain0;
float ain0_value;
float ain0_voltage;
float ain0_temp;

ifp_ain1 = fopen(ain1_in, "r");
if (ifp_ain1 == NULL) {printf("Unable to ain1.\n");}

while(1)
  {
  fseek(ifp_ain0, 0, SEEK_SET);
  fscanf(ifp_ain1, "%f", &ain0_value);
  printf("AIN0 reading [of 1800]: %f\n", ain0_value);
  ain0_voltage = ((ain0_value/1800.0) * 1.8);
  printf("AIN0 voltage [V]: %f\n", ain0_voltage);
  ain0_temp = (ain0_voltage/.010);
  printf("AIN0 temperature [F]: %f\n\n", ain0_temp);
  delay_sec(2.0);
  }
fclose(ifp_ain1);
return 1;
}

//**************************************************************
//function definitions
//**************************************************************

void delay_sec(float delay_value)
{
time_t  now, later;

now = time(NULL);
```

```
later = time (NULL);

while(difftime(later, now) < delay_value)
  {
  later = time(NULL);                //keep checking time
  }
}


//*************************************************************
```

5.9   SERIAL COMMUNICATIONS

Processors must often exchange data with peripheral devices. Data may be exchanged by using parallel or serial techniques. With parallel techniques, an entire byte or word of data is sent simultaneously from the transmitting device to the receiving device. While this is efficient from a time point of view, it requires multiple, parallel lines for the data transfer.

In serial transmission, a byte of data is sent a single bit at a time. Once eight bits have been received at the receiver, the data byte is reconstructed. While this is inefficient from a time point of view, it only requires a line (or two) to transmit the data.

Serial communication techniques provide a vital link between BeagleBone and input devices and output devices. In this section, we investigate the serial communication features beginning with a review of serial communication concepts and terminology. We then investigate serial communication systems available on BeagleBone: the Universal Asynchronous Receiver and Transmitter (UART), the Serial Peripheral Interface (SPI) and networking features. Before discussing the different serial communication features aboard BeagleBone, we review serial communication terminology.

5.9.1   SERIAL COMMUNICATION TERMINOLOGY

In this section, we review common terminology associated with serial communication.

Asynchronous versus Synchronous Serial Transmission: In serial communications, the transmitting and receiving device must be synchronized to one another and use a common data rate and protocol. Synchronization allows both the transmitter and receiver to be expecting data transmission/reception at the same time. There are two basic methods of maintaining "sync" between the transmitter and receiver: asynchronous and synchronous.

In an asynchronous serial communication system, such as the UART, framing bits are used at the beginning and end of a data byte. These framing bits alert the receiver that an incoming data byte has arrived and also signals the completion of the data byte reception. The data rate for an asynchronous serial system is typically much slower than the synchronous system, but it only requires a single wire between the transmitter and receiver.

A synchronous serial communication system maintains "sync" between the transmitter and receiver by employing a common clock between the two devices. Data bits are sent and received on the edge of the clock. This allows data transfer rates higher than with asynchronous techniques but requires two lines, data and clock, to connect the receiver and transmitter.

Baud rate: Data transmission rates are typically specified as a Baud rate or bits per second rate. For example, 9600 Baud indicates the data is being transferred at 9600 bits per second.

Full Duplex: Often serial communication systems must both transmit and receive data. To perform transmission and reception simultaneously requires separate hardware for transmission and reception. A single duplex system has a single complement of hardware that must be switched from transmission to reception configuration. A full duplex serial communication system has separate hardware for transmission and reception.

Non–return to Zero (NRZ) Coding Format: There are many different coding standards used within serial communications. The important point is the transmitter and receiver must use a common coding standard so data may be interpreted correctly at the receiving end. In NRZ coding a logic one is signaled by a logic high during the entire time slot allocated for a single bit; whereas, a logic zero is signaled by a logic low during the entire time slot allocated for a single bit.

The RS–232 Communication Protocol: When serial transmission occurs over a long distance, additional techniques may be used to insure data integrity. Over long distances logic levels degrade and may be corrupted by noise. At the receiving end, it is difficult to discern a logic high from a logic low. The RS–232 standard has been around for some time. With the RS–232 standard (EIA–232), a logic one is represented with a -12 VDC level while a logic zero is represented by a +12 VDC level. Chips are commonly available (e.g., MAX232) that convert the 5 and 0 V output levels from a transmitter to RS–232 compatible levels and convert back to 5V and 0 V levels at the receiver. The RS–232 standard also specifies other features for this communication protocol. The standard specifies several signals including [Horowitz and Hill]:

- **TX:** transmit

- **RX:** receive

- **CTS:** clear to send

- **RTS:** request to send

Depending on the specific peripheral device connected, some or all of these pins will be used in a given application.

The serial communication logic levels from BeagleBone are at 3.3 VDC for logic one and 0 VDC for logic zero. To communicate with an RS–232 device a series of level shifters are required. As shown in Figure 5.20a), the signals from BeagleBone must be first shifted to TTL compatible levels (0 and 5 VDC) and then to RS–232 compatible levels.

Parity: To further enhance data integrity during transmission, parity techniques may be used. Parity is an additional bit (or bits) that may be transmitted with the data byte. With a single parity

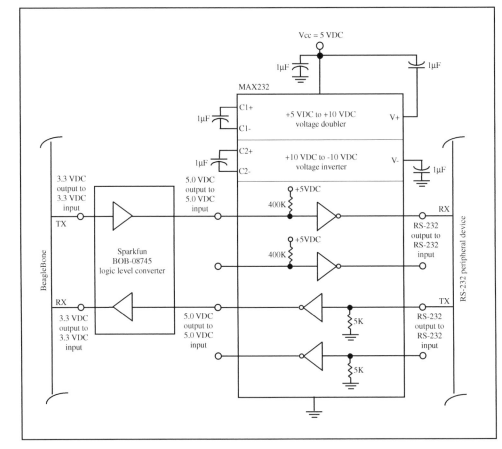

Figure 5.20: Serial communication level shifting.

bit, a single bit error may be detected. Parity may be even or odd. In even parity, the parity bit is set to one or zero such that the number of ones in the data byte including the parity bit is even. In odd parity, the parity bit is set to one or zero such that the number of ones in the data byte including the parity bit is odd. At the receiver, the number of bits within a data byte including the parity bit are counted to insure that parity has not changed, indicating an error, during transmission.

ASCII: The American Standard Code for Information Interchange or ASCII is a standardized, seven bit method of encoding alphanumeric data. It has been in use for many decades, so some of the characters and actions listed in the ASCII table are not in common use today. However, ASCII is still the most common method of encoding alphanumeric data. The ASCII code is provided in Figure 5.21. For example, the capital letter "G" is encoded in ASCII as 0x47. The "0x" symbol indicates the hexadecimal number representation. Unicode is the international counterpart of ASCII. It provides standardized 16-bit encoding format for the written languages of the world.

ASCII is a subset of Unicode. The interested reader is referred to the Unicode home page website, www.unicode.org, for additional information on this standardized encoding format.

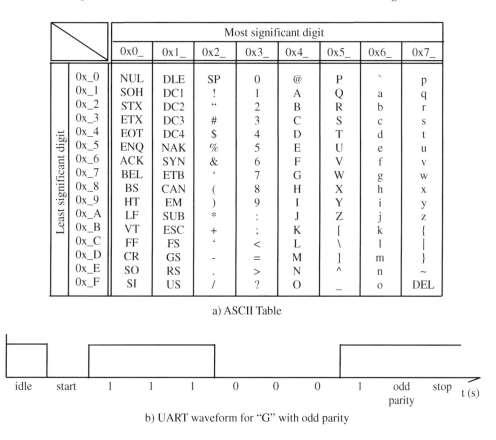

a) ASCII Table

b) UART waveform for "G" with odd parity

Figure 5.21: ASCII Code. The ASCII code is used to encode alphanumeric characters. The "0x" indicates hexadecimal notation in the C programming language [Barrett and Pack].

5.9.2   SERIAL UART

The serial UART (or Universal Asynchronous Receiver and Transmitter) provides for full duplex (two way) communication between a receiver and transmitter. This is accomplished by equipping the processor with independent hardware for the transmitter and receiver. The UART is typically used for asynchronous communication. That is, there is not a common clock between the transmitter and receiver to keep them synchronized with one another. To maintain synchronization between the transmitter and receiver, framing start and stop bits are used at the beginning and end of each data byte in a transmission sequence as shown in Figure 5.21b).

UART channel	serial port	RX	TX	CTS	RTS
UART1	/dev/ttyO1	P9, pin 26 Mux mode: 0	P9, pin 24 Mux mode: 0	P9, pin 20 Mux mode: 0	P9, pin 19 Mux mode: 0
UART2	/dev/ttyO2	P9, pin 22 Mux mode: 1	P9, pin 21 Mux mode: 1	P8, pin 37 Mux mode: 6	P8, pin 38 Mux mode: 6
UART3	/dev/ttyO3	not avail on header	not avail on header	not avail on header	not avail on header
UART4	/dev/ttyO4	P9, pin 11 Mux mode: 6	P9, pin 13 Mux mode: 6	P8, pin 35 Mux mode: 6	P8, pin 33 Mux mode: 6
UART5	/dev/ttyO5	P8, pin 38 Mux mode: 4	P8, pin 37 Mux mode: 4		

Figure 5.22: BeagleBone UART summary.

BeagleBone UART subsystem description–Linux 3.2

BeagleBone is equipped with six UART channels designated UART 0 through UART 5. UART 0 is dedicated to the USB port while UART 1 to 5 are available to the user via the expansion ports as summarized in Figure 5.22. The UART may be configured via the Linux file system. Access to the transmit (TX) and receive (RX) pins are via the following files:

```
uart1_tx: /sys/kernel/debug/omap_mux/uart1_txd
uart1_rx: /sys/kernel/debug/omap_mux/uart1_rxd
uart2_tx: /sys/kernel/debug/omap_mux/spi0_d0
uart2_rx: /sys/kernel/debug/omap_mux/spi0_sclk
uart4_tx: /sys/kernel/debug/omap_mux/gpmc_wpn
uart4_rx: /sys/kernel/debug/omap_mux/gpmc_wait0
uart5_tx: /sys/kernel/debug/omap_mux/lcd_data8
uart5_rx: /sys/kernel/debug/omap_mux/lcd_data9
```

To properly configure the UART system the following parameters must be set:

- baud rate

- character size

- parity type

- flow control type

UART support functions in C –Linux 3.2

The UART features are usually not accessed directly but instead through a series of support functions via an Applications Programming Interface (API). We use the termios API in the following example.

Example: In this example UART1 is configured for transmission at 9600 BAUD and continuously communicates the ASCII character "G."

```
//*************************************************************************
//uart1.cpp - configures BeagleBone uart1 for tranmission and 9600 Baud
//and repeatedly sends the character G via uart1 tx pin (P9, 24)
//*************************************************************************

#include <stdio.h>
#include <stddef.h>
#include <time.h>
#include <iostream>
#include <termios.h>
#include <fcntl.h>
#include <unistd.h>
#include <sys/types.h>
#include <string.h>

int main(void)
{
//define file handle for uart1
FILE *ofp_uart1_tx, *ofp_uart1_rx;

//define mux mode for uart1 tx
int mux_mode_uart1_tx = 0X8000;
int mux_mode_uart1_rx = 0X8020;

//configure uart1 for transmission
ofp_uart1_tx = fopen("/sys/kernel/debug/omap_mux/uart1_txd", "w");
if(ofp_uart1_tx == NULL) printf("Unable to open uart1 tx file.\n");
fseek(ofp_uart1_tx, 0, SEEK_SET);
fprintf(ofp_uart1_tx, "0x%2x", mux_mode_uart1_tx);
fflush(ofp_uart1_tx);

//configure uart1 for reception
ofp_uart1_rx = fopen("/sys/kernel/debug/omap_mux/uart1_rxd", "w");
if(ofp_uart1_rx == NULL) printf("Unable to open uart1 rx file.\n");
```

```c
fseek(ofp_uart1_rx, 0, SEEK_SET);
fprintf(ofp_uart1_rx, "0x%2x", mux_mode_uart1_rx);
fflush(ofp_uart1_rx);

//uart1 configuration using termios
termios uart1;
int fd;

//open uart1 for tx/rx, not controlling device
if((fd = open("/dev/tty01", O_RDWR | O_NOCTTY)) < 0)
  printf("Unable to open uart1 access.\n");

//get attributes of uart1
if(tcgetattr(fd, &uart1) < 0)
  printf("Could not get attributes of UART1 at tty01\n");

//set Baud rate
if(cfsetospeed(&uart1, B9600) < 0)
  printf("Could not set baud rate\n");
else
  printf("Baud rate: 9600\n");

//set attributes of uart1
uart1.c_iflag = 0;
uart1.c_oflag = 0;
uart1.c_lflag = 0;
tcsetattr(fd, TCSANOW, &uart1);

char byte_out[] = {0x47};

//set ASCII charater G repeatedly
while(1)
  {
  write(fd, byte_out, strlen(byte_out)+1);
  }

close(fd);
}
```

```
//**********************************************************
```

UART support dts files in C –Linux 3.8

uart1.dts: All of the other files are for reference with outputs included in the recent software images already. This one is an example that shows how you can enable the pinmux for UART1 using bone–pinmux–helper and enable the UART as well.

```
//***********************************************************

/*
 * Copyright (C) 2013 CircuitCo
 * Copyright (C) 2013 Texas Instruments
 *
 * This program is free software; you can redistribute it and/or modify
 * it under the terms of the GNU General Public License version 2 as
 * published by the Free Software Foundation.
 */
/dts-v1/;
/plugin/;

/ {
    compatible = "ti,beaglebone", "ti,beaglebone-black";

    /* identification */
    part-number = "uart1";
    version = "00A0";

    fragment@0 {
            target = <&am33xx_pinmux>;
            __overlay__ {
                    pinmux_serial1: pinmux_serial1_pins {
                            pinctrl-single,pins = <
                0x184  0x20 /*P9_24(ZCZ ball D15) RX-enabled MODE 0*/
                0x180  0x20 /*P9_26(ZCZ ball D16) RX-enabled MODE 0*/
            >;
                    };
            };
    };

    fragment@1 {
```

```
                    target = <&ocp>;
                    __overlay__ {
            serial1_pinmux_helper {
                compatible  = "bone-pinmux-helper";
                status      = "okay";
                pinctrl-names = "default";
                        pinctrl-0 = <&pinmux_serial1>;
                        };
                };
        };

        fragment@2 {
                target = <&uart2>;              /* really uart1 */
                __overlay__ {
                        status  = "okay";
                };
        };
};


//***************************************************************

//*********************************************************************
//uart1.cpp - configures BeagleBone uart1 for tranmission and 9600 Baud
//and repeatedly sends the character G via uart1 tx pin (P9, 24)
//
//To configure the UART first do:
// # dtc -O dtb -o uart1-00A0.dtbo -b 0 -@ uart1.dts
// # cp uart1-00A0.dtbo /lib/firmware/uart1-00A0.dtbo
// # echo uart1 > /sys/devices/bone_capemgr.8/slots
//
//Check that the pinmux has been configured via:
// # cat /sys/kernel/debug/pinctrl/44e10800.pinmux/pinmux-pins
// pin 96 (44e10980): serial1_pinmux_helper.14 (GPIO UNCLAIMED)
//function pinmux_serial1_pins group pinmux_serial1_pins
// pin 97 (44e10984): serial1_pinmux_helper.14 (GPIO UNCLAIMED)
// function pinmux_serial1_pins group pinmux_serial1_pins
//
//*********************************************************************

#include <stdio.h>
```

```
#include <stddef.h>
#include <time.h>
#include <iostream>
#include <termios.h>
#include <fcntl.h>
#include <unistd.h>
#include <sys/types.h>
#include <string.h>

int main(void)
{
//define file handle for uart1
FILE *ofp_uart1_tx, *ofp_uart1_rx;

//uart1 configuration using termios
termios uart1;
int fd;

//open uart1 for tx/rx, not controlling device
if((fd = open("/dev/ttyO1", O_RDWR | O_NOCTTY)) < 0)
  printf("Unable to open uart1 access.\n");

//get attributes of uart1
if(tcgetattr(fd, &uart1) < 0)
  printf("Could not get attributes of UART1 at ttyO1\n");

//set Baud rate
if(cfsetospeed(&uart1, B9600) < 0)
  printf("Could not set baud rate\n");
else
  printf("Baud rate: 9600\n");

//set attributes of uart1
uart1.c_iflag = 0;
uart1.c_oflag = 0;
uart1.c_lflag = 0;
tcsetattr(fd, TCSANOW, &uart1);

char byte_out[] = {0x47};
```

```
//set ASCII charater G repeatedly
while(1)
  {
  write(fd, byte_out, strlen(byte_out)+1);
  }

close(fd);
}


//************************************************************
```

BeagleBone may be equipped with RS–232 compatible features using the BeagleBone RS232 Cape. Full documentation and software support is available for the Cape [circuitco].

5.9.3   SERIAL PERIPHERAL INTERFACE (SPI)

The Serial Peripheral Interface or SPI also provides for two–way serial communication between a transmitter and a receiver. In the SPI system, the transmitter and receiver share a common clock source. This requires an additional clock line between the transmitter and receiver but allows for higher data transmission rates as compared to the USART. The SPI system allows for fast and efficient data exchange between microcontrollers or peripheral devices. There are many SPI compatible external systems available to extend the features of the microcontroller. For example, a liquid crystal display or a multi–channel digital–to–analog converter could be added to the processor using the SPI system.

SPI Operation

The SPI may be viewed as a synchronous 16–bit shift register with an 8–bit half residing in the transmitter and the other 8–bit half residing in the receiver as shown in Figure 5.23. The transmitter is designated the master since it is providing the synchronizing clock source between the transmitter and the receiver. The receiver is designated as the slave. A slave is chosen for reception by taking its Slave Select line low. When the line is taken low, the slave's shifting capability is enabled.

SPI transmission is initiated by loading a data byte into the master configured SPI Data Register. At that time, the SPI clock generator provides clock pulses to the master and also to the slave via the serial clock pin. A single bit is shifted out of the master designated shift register on the Master Out Slave In (MOSI) processor pin on every serial clock pulse. The data is received at the MOSI pin of the slave designated device. At the same time, a single bit is shifted out of the Master In Slave Out (MISO) pin of the slave device and into the MISO pin of the master device. After eight master serial clock pulses, a byte of data has been exchanged between the master and slave designated SPI devices. The serial transmission does not have to be bi–directional. In these applications the return line from the slave to the master device is not connected.

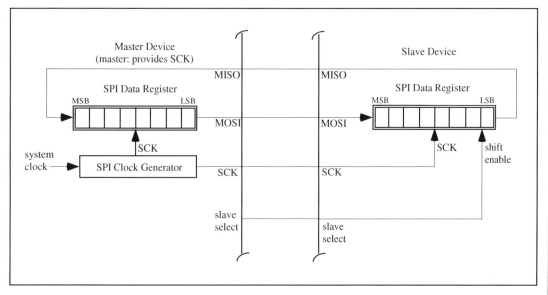

Figure 5.23: SPI Overview.

Bonescript SPI support

An SPI signal may be generated on many different BeagleBone pins using the Bonescript function shiftOut. The format of the function is:

```
shiftOut(dataPin, clockPin, bitOrder, value);
```

The bitOrder may be set for MSBFIRST (most significant bit first) or LSBFIRST (least significant bit first). In the example below the value 0xAC is continuously sent out via BeagleBone P8, pin 3. The SPI clock signal is sent via P8, pin 11. The bit rate in this example is approximately 1,500 bits per second. The SPI data and clock signals may be examined using an oscilloscope.

```
//**********************************************************

var b = require('bonescript');

//Work with old and new bonescript
var bone = (typeof b.bone.pins != 'undefined') ? b.bone.pins : b.bone;
setup = function() {};

var spi_data_pin = bone.P8_13;
var spi_clk_pin = bone.P8_11;
```

```
b.pinMode(spi_data_pin, b.OUTPUT);
b.pinMode(spi_clk_pin, b.OUTPUT);
b.shiftOut(spi_data_pin, spi_clk_pin, b.LSBFIRST, 0xAC);


//***********************************************************
```

BeagleBone SPI features

For additional flexibility and increased data rate, BeagleBone dedicated SPI features may be used. BeagleBone is equipped with two SPI channels designated SPIO0 and SPI1. The SPI features may be accessed via the header pins as shown in Figure 5.24.

BeagleBone Serial Peripheral Interface (SPI) features		
Header pin	**Signal name**	**Mode**
SPIO0		
P9.17	spi0_cs0	0
P9.18	spi0_d1	0
P9.21	spi0_d0	0
P9.22	spi0_sclk	0
SPI1		
P9.28	spi1_cs0	3
P9.29	spi1_d0	3
P9.30	spi1_d1	3
P9.31	spi1_sclk	3

Figure 5.24: BeagleBone SPI features.

SP support functions in C –Linux 3.2

BeagleBone provides considerable support for the SPI system via:

```
/dev/spidev2.0
```

There is considerable additional information on this topic and C support functions available through the Linux Documentation Project.

5.10   PRECISION TIMING

Processors may be used to accomplish time related tasks including generating precision digital signals, generating pulse widths of a specific duration or measuring the parameters of an incoming signal.

Also, pulse width modulation (PWM) techniques may be used to vary the speed of a motor or control specialized motors such as a servo motor. In this section we describe some of the timing features available on BeagleBone. We begin with a review of timing related terminology followed by a review of some of the BeagleBone timing features. We conclude with several timing related examples. In the next section we discuss BeagleBone's PWM features.

5.10.1 TIMING RELATED TERMINOLOGY

In this section, we review timing related terminology including frequency, period, and duty cycle.

Frequency: Consider a signal $x(t)$ that repeats itself. We call this signal periodic with period T, if it satisfies

$$x(t) = x(t + T).$$

To measure the frequency of a periodic signal, we count the number of times a particular event repeats within a one second period. The unit of frequency is Hertz or cycles per second. For example, a sinusoidal signal with a 60 Hz frequency means that a full cycle of a sinusoid signal repeats itself 60 times each second or every 16.67 ms. The period of the signal is 16.67 ms.

Period: The reciprocal of frequency is the period of a waveform. If an event occurs with a rate of 1 Hz, the period of that event is 1 second. To find a period, given the frequency of a signal, or vice versa, we simply need to remember their inverse relationship $f = \frac{1}{T}$ where f and T represent a frequency and the corresponding period, respectively.

Duty Cycle: In many applications, periodic pulses are used as control signals. A good example is the use of a periodic pulse to control a servo motor. To control the direction and sometimes the speed of a motor, a periodic pulse signal with a changing duty cycle over time is used. The periodic pulse signal shown in Figure 5.25(a) is on for 50 percent of the signal period and off for the rest of the period. The pulse shown in (b) is on for only 25 percent of the same period as the signal in (a) and off for 75 percent of the period. The duty cycle is defined as the percentage of the period a signal is on or logic high. Therefore, we call the signal in Figure 5.25(a) as a periodic pulse signal with a 50 percent duty cycle and the corresponding signal in (b), a periodic pulse signal with a 25 percent duty cycle. These features are discussed in more detail in the PWM section.

5.10.2 BEAGLEBONE TIMING CAPABILITY SYSTEM–LINUX 3.2

Depending on version and power supply source, BeagleBone is clocked from 500 to 1 GHz. This allows BeagleBone to measure the characteristics of high frequency input signals or generate high frequency digital signals. We limit our discussion to the timing functions available in C and also the Linux system.

Earlier in the chapter we provided an example to blink an LED at 5 second intervals (led3.cpp). In this example we used the timing features available within the C programming language. The "time" function in ANSI C returns the current calendar time in seconds that have elapsed since January 1, 1970. Time hacks may be taken at different times for use in delay functions or to measure intervals

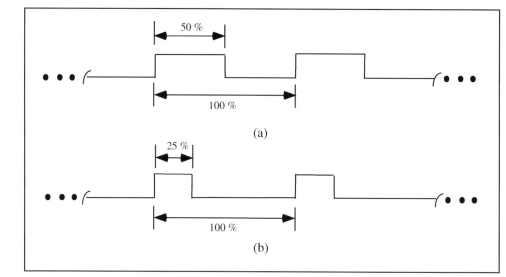

Figure 5.25: Two signals with the same period but different duty cycles. The top figure (a) shows a periodic signal with a 50% duty cycle and the lower figure (b) displays a periodic signal with a 25% duty cycle [Barrett and Pack].

with the resolution of seconds. The ANSI C library also provides a "difftime" function that provides the difference in time between two time hacks. The difference is in seconds as a double type variable [Kelley and Pohl]. The <time.h> header file must be included to use these features.

To achieve better time resolution the Linux "gettimeofday" function may be used. The function returns the current time in seconds and microseconds in a timeval structure since January 1, 1970.

```
struct timeval{
   time_t        tv_sec;      //seconds
   suseconds_t   tv_usec;     //microseconds
};
```

To use these features the <sys/time.h> header file must be included. In the following example, the "gettimeofday" function is used to generate a 100 Hz, 50% duty cycle signal on header P8, pin 3 (GPIO1_6 designated as gpio38). The signal may be easily changed to 100 kHz signal simply by changing the argument of the "delay_us" from 5000 to 5.

BeagleBone Original Linux–3.2

```
//****************************************************************
//BeagleBone Original, Linux 3.2
//sq_wave:  generates a 100 Hz, 50% duty cycle signal on header
//P8, pin 3 (GPIO1_6 designated as gpio38).
```

```
//*****************************************************************

#include <stdio.h>
#include <stddef.h>
#include <time.h>
#include <sys/time.h>

#define  output "out"
#define  input  "in"

void delay_us(int);

int  main (void)
{
//define file handles
FILE *ofp_gpm6_ad6, *ofp_export, *ofp_gpio38_value, *ofp_gpio38_direction;

//define pin variables
int mux_mode = 0x007, pin_number = 38, logic_status = 1;
char* pin_direction = output;

ofp_gpm6_ad6 = fopen("/sys/kernel/debug/omap_mux/gpmc_ad6", "w");
if(ofp_gpm6_ad6 == NULL) {printf("Unable to open gpmc_ad6.\n");}
fseek(ofp_gpm6_ad6, 0, SEEK_SET);
fprintf(ofp_gpm6_ad6, "0x%02x", mux_mode);
fflush(ofp_gpm6_ad6);

ofp_export = fopen("/sys/class/gpio/export", "w");
if(ofp_export == NULL) {printf("Unable to open export.\n");}
fseek(ofp_export, 0, SEEK_SET);
fprintf(ofp_export, "%d", pin_number);
fflush(ofp_export);

ofp_gpio38_direction = fopen("/sys/class/gpio/gpio38/direction", "w");
if(ofp_gpio38_direction==NULL){printf("Unable to open gpio38_direction.\n");}
fseek(ofp_gpio38_direction, 0, SEEK_SET);
fprintf(ofp_gpio38_direction, "%s",  pin_direction);
fflush(ofp_gpio38_direction);
```

```
ofp_gpio38_value = fopen("/sys/class/gpio/gpio38/value", "w");
if(ofp_gpio38_value == NULL) {printf("Unable to open gpio38_value.\n");}
fseek(ofp_gpio38_value, 0, SEEK_SET);
logic_status = 1;
fprintf(ofp_gpio38_value, "%d", logic_status);
fflush(ofp_gpio38_value);


while(1)
  {
  delay_us(5000);
  if(logic_status == 1) logic_status = 0;
    else logic_status = 1;
  //write to gpio38
  fprintf(ofp_gpio38_value, "%d", logic_status);
  fflush(ofp_gpio38_value);
  }
fclose(ofp_gpm6_ad6);
fclose(ofp_export);
fclose(ofp_gpio38_direction);
fclose(ofp_gpio38_value);
return 1;
}


//****************************************************************

void delay_us(int desired_delay_us)
{
struct timeval  tv_start;  //start time hack
struct timeval  tv_now;    //current time hack
int elapsed_time_us;

gettimeofday(&tv_start, NULL);
elapsed_time_us = 0;

while(elapsed_time_us <  desired_delay_us)
  {
  gettimeofday(&tv_now, NULL);
```

```
  if(tv_now.tv_usec >= tv_start.tv_usec)
    elapsed_time_us = tv_now.tv_usec - tv_start.tv_usec;
  else
    elapsed_time_us = (1000000 - tv_start.tv_usec) + tv_now.tv_usec;
  //printf("start: %ld \n", tv_start.tv_usec);
  //printf("now: %ld \n", tv_now.tv_usec);
  //printf("desired: %d \n", desired_delay_ms);
  //printf("elapsed: %d \n\n", elapsed_time_ms);
  }
}


//*********************************************************************
     BeagleBone Black Linux-3.8
//*********************************************************************
//sq_wave:  generates a 100 Hz, 50% duty cycle signal on header
//P8, pin 13 (GPIO1_12 designated as gpio44).
//*********************************************************************

#include <stdio.h>
#include <stddef.h>
#include <time.h>
#include <sys/time.h>

#define  output "out"
#define  input  "in"

void delay_us(int);

int  main (void)
{
//define file handles
FILE *ofp_export, *ofp_gpio44_value, *ofp_gpio44_direction;

//define pin variables
int pin_number = 44, logic_status = 1;
char* pin_direction = output;

ofp_export = fopen("/sys/class/gpio/export", "w");
if(ofp_export == NULL) {printf("Unable to open export.\n");}
```

```c
fseek(ofp_export, 0, SEEK_SET);
fprintf(ofp_export, "%d", pin_number);
fflush(ofp_export);

ofp_gpio44_direction = fopen("/sys/class/gpio/gpio44/direction", "w");
if(ofp_gpio44_direction == NULL) {printf("Unable to open gpio44_direction.\n");}
fseek(ofp_gpio44_direction, 0, SEEK_SET);
fprintf(ofp_gpio44_direction, "%s",  pin_direction);
fflush(ofp_gpio44_direction);

ofp_gpio44_value = fopen("/sys/class/gpio/gpio44/value", "w");
if(ofp_gpio44_value == NULL) {printf("Unable to open gpio44_value.\n");}
fseek(ofp_gpio44_value, 0, SEEK_SET);
logic_status = 1;
fprintf(ofp_gpio44_value, "%d", logic_status);
fflush(ofp_gpio44_value);


while(1)
  {
  delay_us(5000);
  if(logic_status == 1) logic_status = 0;
    else logic_status = 1;
  //write to gpio44
  fprintf(ofp_gpio44_value, "%d", logic_status);
  fflush(ofp_gpio44_value);
  }
fclose(ofp_export);
fclose(ofp_gpio44_direction);
fclose(ofp_gpio44_value);
return 1;
}

//****************************************************************

void delay_us(int desired_delay_us)
{
struct timeval  tv_start;  //start time hack
struct timeval  tv_now;    //current time hack
```

```
int elapsed_time_us;

gettimeofday(&tv_start, NULL);
elapsed_time_us = 0;

while(elapsed_time_us <  desired_delay_us)
  {
  gettimeofday(&tv_now, NULL);
  if(tv_now.tv_usec >= tv_start.tv_usec)
    elapsed_time_us = tv_now.tv_usec - tv_start.tv_usec;
  else
    elapsed_time_us = (1000000 - tv_start.tv_usec) + tv_now.tv_usec;
  //printf("start: %ld \n", tv_start.tv_usec);
  //printf("now: %ld \n", tv_now.tv_usec);
  //printf("desired: %d \n", desired_delay_ms);
  //printf("elapsed: %d \n\n", elapsed_time_ms);
  }
}


//********************************************************************
```

5.11   PULSE WIDTH MODULATION (PWM)

In this section, we discuss a method to control the speed of a DC motor using a pulse width modulated (PWM) signal. If we turn on a DC motor and provide the required voltage, the motor will run at its maximum speed. Suppose we turn the motor on and off rapidly, by applying a periodic signal. The motor at some point can not react fast enough to the changes of the voltage values and will run at the speed proportional to the average time the motor was turned on. By changing the duty cycle, we can control the speed of a DC motor as we desire. Suppose again we want to generate a speed profile shown in Figure 5.26. As shown in the figure, we want to accelerate the speed, maintain the speed, and decelerate the speed for a fixed amount of time.

As an example, for passenger comfort, an elevator control system does not immediately operate the elevator motor at full speed. The elevator motor speed will ramp up gradually from stop to desired speed. As the elevator approaches, the desired floor it will gradually ramp back down to stop.

Earlier in the chapter we discussed the signal parameters of frequency and duty cycle. A PWM signal maintains a constant baseline frequency. The duty cycle of the signal is varied as required by the specific application. Also, the polarity of the signal may be active high or active low.

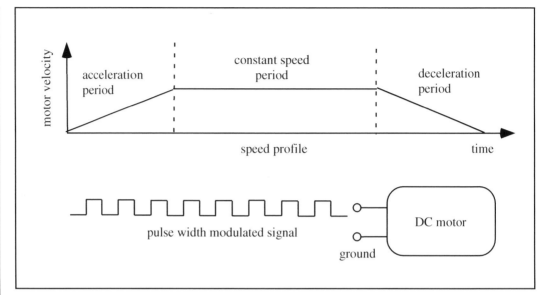

Figure 5.26: Speed profile of a DC motor over time with a PWM signal applied [Barrett and Pack].

5.11.1 BEAGLEBONE PWM SUBSYSTEM (PWMSS) DESCRIPTION

The description provided here was adapted from the "AM335X PWMSS Driver's Guide [AM335x]." BeagleBone is equipped the PWMSS system. It is subdivided into the:

- enhanced high resolution PWM (eHRPWM) system

- enhanced Captured (eCAP) system

- enhanced Quadrature Encoded Puse (eQEP) system

Due to space limitations we only discuss the eHRPWM system in detail. The eHRPWM system is supported by 16–bit timers for period and frequency. The eHRPWM system consists of two instances of two channels each. The instances and channels are designated as:

`ehrpwm.i:j`

where i is the instance and j (0 or 1) is the channel.

For example, in BeagleBone, EHRPWM1A (P9, pin 14) is designated ehrpwm.1:0 and EHRPWM1B (P9, pin 16) is designated ehrpwm.1:1.

To configure and use the PWM system a four step process is followed.

1. Configure the PWM pin for output.

2. Request the PWM device.

3. Configure the PWM device.

4. Start (and Stop) the PWM device.

The steps are accomplished using the BeagleBone Linux file system. We provide the Linux commands to accomplish each step.

5.11.2 PWM CONFIGURATION–LINUX 3.2

Step 1. Configure the PWM pin for output.

```
>echo 6 > /sys/kernel/debug/omap_mux/gpmc_a2
```

Step 2. Request the PWM device.

```
>echo 1 > /sys/class/pwm/ehrpwm.1:0/request
```

Step 3. Configure the PWM device.

```
>echo 100 > /sys/class/pwm/ehrpwm.1:0/period_freq
>echo 50 > /sys/class/pwm/ehrpwm.1:0/duty_percent
>echo 1 > /sys/class/pwm/ehrpwm.1:0/polarity
```

Step 4. Start (1) (and Stop (0)) the PWM device.

```
>echo 1 > /sys/class/pwm/ehrpwm.1:0/run
```

5.11.3 PWM C SUPPORT FUNCTIONS–LINUX 3.2

In this example we configure EHRPWM1A for 100 Hz and 50% duty cycle.

```
//*************************************************************************
#include <stdio.h>
#include <stddef.h>
#include <time.h>

#define  output "out"
#define  input  "in"

int  main (void)
{
//define file handles
FILE *ofp_gpmc_a2, *pwm_freq, *pwm_req, *pwm_duty, *pwm_polarity, *pwm_run;

//define pin variables
int mux_mode = 0x006, request = 1, freq = 100, duty = 50, polarity=1, run=1;
```

```
ofp_gpmc_a2 = fopen("/sys/kernel/debug/omap_mux/gpmc_a2", "w");
if(ofp_gpmc_a2 == NULL) {printf("Unable to open gpmc_a2.\n");}
fseek(ofp_gpmc_a2, 0, SEEK_SET);
fprintf(ofp_gpmc_a2, "0x%02x", mux_mode);
fflush(ofp_gpmc_a2);

pwm_req = fopen("/sys/class/pwm/ehrpwm.1:0/request", "w");
if(pwm_req == NULL) {printf("Unable to open pwm request.\n");}
fseek(pwm_req, 0, SEEK_SET);
fprintf(pwm_req, "%d", request);
fflush(pwm_req);

pwm_freq = fopen("/sys/class/pwm/ehrpwm.1:0/period_freq", "w");
if(pwm_freq == NULL) {printf("Unable to open pwm frequency.\n");}
fseek(pwm_freq, 0, SEEK_SET);
fprintf(pwm_freq, "%d", freq);
fflush(pwm_freq);

pwm_duty = fopen("/sys/class/pwm/ehrpwm.1:0/duty_percent", "w");
if(pwm_duty == NULL) {printf("Unable to open pwm duty cycle.\n");}
fseek(pwm_duty, 0, SEEK_SET);
fprintf(pwm_duty, "%d", duty);
fflush(pwm_duty);

pwm_polarity = fopen("/sys/class/pwm/ehrpwm.1:0/polarity", "w");
if(pwm_polarity == NULL) {printf("Unable to open pwm polarity.\n");}
fseek(pwm_polarity, 0, SEEK_SET);
fprintf(pwm_polarity, "%d", polarity);
fflush(pwm_polarity);

pwm_run = fopen("/sys/class/pwm/ehrpwm.1:0/run", "w");
if(pwm_run == NULL) {printf("Unable to open pwm run.\n");}
fseek(pwm_run, 0, SEEK_SET);
fprintf(pwm_run, "%d", run);
fflush(pwm_run);

while(1)
  {
```

```
    :


  }

fclose(ofp_gpmc_a2);
fclose(pwm_req);
fclose(pwm_freq);
fclose(pwm_duty);
fclose(pwm_polarity);
fclose(pwm_run);

return 1;
}
//************************************************************************
```

5.11.4 PWM C SUPPORT FUNCTIONS–LINUX 3.8

```
//************************************************************************
//pwm.cpp
//
//This uses PWM to output a "fade" effect to P9_14
//
//Before use, you must:
// echo bone_pwm_P9_14 > /sys/devices/bone_capemgr.*/slots
//
//The basic steps are the same, but an initial configuration
//is provided by the devicetree fragment.
The 'pwm_test'
//driver that provides the file system interface has changed
//and now provides 'period' instead of 'freq' and 'duty' instead
//of 'duty_percent'.  'request' is no longer used.
//
//Note that in /sys/devices/ocp.2/pwm_test_P9_14.12
// the .2 after ocp and the .12 after pwm_test_P9_14 both might change
//************************************************************************
#include <stdio.h>
#include <stddef.h>
#include <time.h>
```

```c
#define   output  "out"
#define   input   "in"

int  main (void)
{
//define file handles
FILE *pwm_period, *pwm_duty, *pwm_polarity, *pwm_run;

//define pin variables
int period = 500000, duty = 250000, polarity = 1, run = 1;
int increment = 1;

pwm_period = fopen("/sys/devices/ocp.2/pwm_test_P9_14.12/period", "w");
if(pwm_period == NULL) {printf("Unable to open pwm period.\n");}
fseek(pwm_period, 0, SEEK_SET);
fprintf(pwm_period, "%d", period);
fflush(pwm_period);

pwm_duty = fopen("/sys/devices/ocp.2/pwm_test_P9_14.12/duty", "w");
if(pwm_duty == NULL) {printf("Unable to open pwm duty cycle.\n");}
fseek(pwm_duty, 0, SEEK_SET);
fprintf(pwm_duty, "%d", duty);
fflush(pwm_duty);

pwm_polarity = fopen("/sys/devices/ocp.2/pwm_test_P9_14.12/polarity", "w");
if(pwm_polarity == NULL) {printf("Unable to open pwm polarity.\n");}
fseek(pwm_polarity, 0, SEEK_SET);
fprintf(pwm_polarity, "%d", polarity);
fflush(pwm_polarity);

pwm_run = fopen("/sys/devices/ocp.2/pwm_test_P9_14.12/run", "w");
if(pwm_run == NULL) {printf("Unable to open pwm run.\n");}
fseek(pwm_run, 0, SEEK_SET);
fprintf(pwm_run, "%d", run);
fflush(pwm_run);

while(1)
  {
if(duty >= period) increment = -1;
```

```
else if(duty <= 0) increment = 1;
duty += increment;
fseek(pwm_duty, 0, SEEK_SET);
fprintf(pwm_duty, "%d", duty);
fflush(pwm_duty);
  }

fclose(pwm_period);
fclose(pwm_duty);
fclose(pwm_polarity);
fclose(pwm_run);

return 1;
}
//************************************************************************
```

5.12   NETWORKING

BeagleBone is equipped with several networking systems the I2C for small area (circuit board) networking, the Controller Area Network (CAN) for system level networking and the Ethernet 10/100 PHY for local area networks (LAN). We discuss each system in turn.

5.12.1   INTER–INTEGRATED CIRCUIT (I2C) BUS

The Inter–IC bus (I2C) provides a method to interconnect multiple system components together residing in a small, circuit board size area. The I2C system is also referred to as the IIC bus or the two–wire interface (TWI). The I2C system consists of a two wire 100k bps (bit per second) bus. The 100k bps bus speed is termed the standard mode but the bus may also operate at higher data rates. There are multiple I2C compatible peripheral components (e.g., LCD displays, sensors, etc.) [I2C].

A large number of devices (termed nodes) may be connected to the I2C bus. The I2C system uses a standard protocol to allow the nodes to send and receive data from the other devices. All nodes on the bus are assigned a unique 7–bit address. The eighth bit of the address register is used to specify the operation to be performed (read or write). Additional devices may be added to the I2C based system as it evolves [I2C].

The basic I2C bus architecture is shown in Figure 5.27. The two wire bus consists of the serial clock line (SCL) and the serial data line (SDA). These lines are pulled up to logic high by the SCL and the SDA pull up resistors. Nodes connected to the bus can drive either of the bus lines to ground (logic 0). Devices within an I2C bus configuration must share a common ground [I2C].

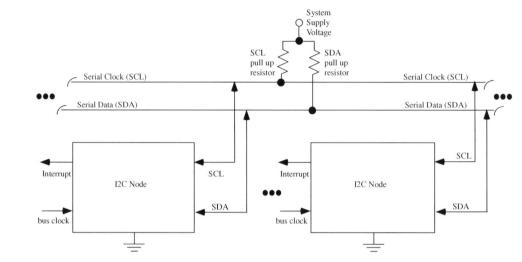

Figure 5.27: I2C configuration.

BeagleBone I2C subsystem description

BeagleBone is equipped with two I2C channels designated I2C1 and I2C2. The I2C2 channel is dedicated for EEPROM use and should not be tampered with. The I2C1 channel is available at P9 pin 17 for I2C1_SCL (Mode 2) and P9 pin 18 for I2C1_SDA (Mode 2). There is considerable additional information on this topic and C support functions available through the Linux Documentation Project [Coley].

5.12.2   CONTROLLER AREA NETWORK (CAN) BUS

The Controller Area Network or CAN bus was originally developed for the automotive industry in the 1980's. There are two different CAN protocols: A the basic or standard version and B the extended or full version. The CAN protocol allows a number of processors or nodes to be linked via a twisted pair cable. The nodes may exchange data serially at up to 1 Mbit/second data rates. Each node on the CAN bus can serve as the master and can send or receive data messages over the bus [COMSOL].

The CAN message format consists of four different types of frames: data, remote, error and overload. The data frame is shown in Figure 5.28. Each frame consists of a series of fields. Embedded in the message is an identifier. For an incoming frame, each node will examine the identifier to determine if the frame is intended for the node. This allows a message to be sent to a specific node or group of nodes [COMSOL].

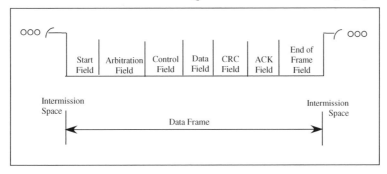

Figure 5.28: CAN data frame [COMSOL].

BeagleBone CAN subsystem description
BeagleBone may be equipped with CAN features using either the TT3201 CAN Cape or the BeagleBone CAN Bus Cape. Each Cape is supported with full documentation and software support [Coley, circuitco].

5.12.3 ETHERNET

BeagleBone is equipped with 10/100 Ethernet capability via the SMSC LAN8710A integrated circuit to provide local area network (LAN) capability. This chip implements the Media Independent Interface (MII) physical layer (PHY) of the Open Systems Interconnection (OSI) model. The PHY implements the hardware send and receive protocol by breaking a serial data stream into frames. The PHY is configured for auto negotiation which allows two connected devices to choose common transmission parameters [Coley, SMSC].

5.13   LIQUID CRYSTAL DISPLAY (LCD) INTERFACE

BeagleBone provides full support for both 24 and 16–bit LCD interfaces. A 7 inch LCD Cape is available from Circuitco. The TFT LCD has 800 by 480 pixel resolution and also supports touch screen features. The Cape provides an easy to use 39–pin interface for the large LCD display. Also, Linux Ångström distribution images after 6.18.12 directly support the Cape. Full support data for the LCD is provided in "BeagleBone LCD7 Cape Rev A3 System Reference Manual."

If an application can not support use of the 39–pin Cape interface and other desired subsystems, a smaller footprint LCD may be employed. Serial LCDs employing a UART or SPI interface are readily available. In the following example, we equip BeagleBone with a two line, 16 character display employing a parallel interface.

Example: In this example we equip BeagleBone with the Sparkfun LCD–09052 16 by 2 character LCD. This is a 3.3 VDC LCD with White on black characters. The interface between BeagleBone and the LCD is provided in Figure 5.29.

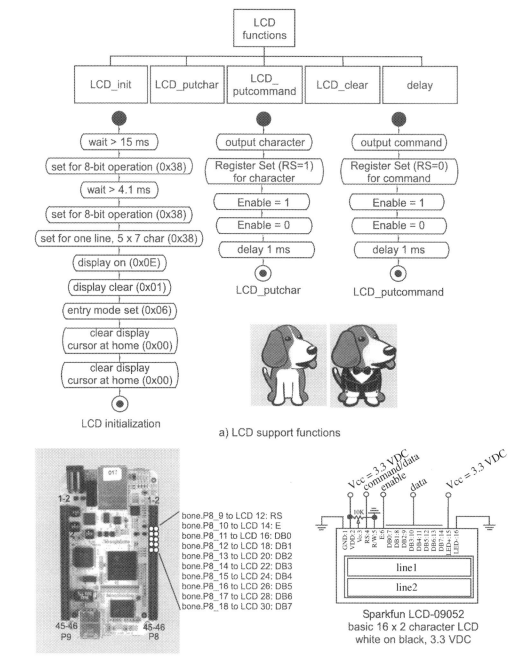

a) LCD support functions

b) LCD hardware interface

Figure 5.29: BeagleBone LCD interface.

5.13.1 C SUPPORT FUNCTIONS

The C code for the LCD interface and support functions are provided in Appendix B.

5.14   INTERRUPTS

A processor normally executes instructions in an orderly fetch–decode–execute sequence as dictated by a user–written program as shown in Figure 5.30. However, the processor must be equipped to handle unscheduled (although planned), higher priority events that might occur inside or outside the processor. To process such events, a processor requires an interrupt system.

The interrupt system onboard a processor allows it to respond to higher priority events. Appropriate responses to these events may be planned, but we do not know when these events will occur. When an interrupt event occurs, the processor will normally complete the instruction it is currently executing and then transition program control to interrupt event specific tasks. These tasks, which resolve the interrupt event, are organized into a function called an interrupt service routine (ISR). Each interrupt will normally have its own interrupt specific ISR. Once the ISR is complete, the processor will resume processing where it left off before the interrupt event occurred.

5.14.1 BONESCRIPT INTERRUPT SUPPORT

Bonescript provides an "attachInterrupt" function to support interrupt processing on Beaglebone. The function associates an interrupt event with the desired actions to be accomplished when the event occurs. The function has three arguments: the pin to monitor for the interrupt event, the type of pin activity to initiate the interrupt, and the name of the interrupt service routine (ISR) to be executed when the interrupt event occurs.

Example: In this example a tactile switch is connected to P8, pin 15. When P8, pin 15 experiences a rising edge the ISR is executed. In the main loop of the program a green LED is blinking at 100 ms intervals. When the switch is depressed and released, creating a rising edge on P8, pin 15 the ISR is executed which blinks a red LED at 50 ms intervals. A circuit diagram for this example is provided in Figure 5.31.

```
//************************************************************
var b = require('bonescript');

//Work with old and new bonescript
var bone = (typeof b.bone.pins != 'undefined') ? b.bone.pins : b.bone;
setup = function() {};

var greenLED = bone.P8_13;
var redLED = bone.P9_14;
var inputSW = bone.P8_15;
var main_delay = 100;
```

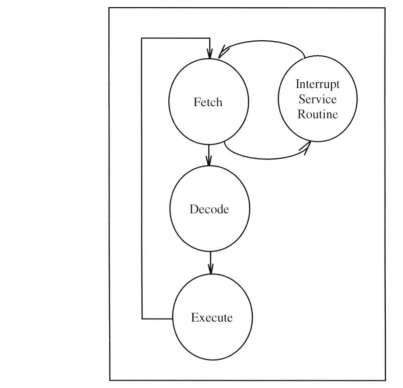

Figure 5.30: Processor interrupt response.

```
var inter_delay = 50;
var state = b.LOW;

b.pinMode(greenLED, b.OUTPUT);
b.pinMode(redLED, b.OUTPUT);
b.pinMode(inputSW, b.INPUT);
b.attachInterrupt(inputSW, RISING, blink_red);

function loop()
    {
    state = (state == b.HIGH) ? b.LOW : b.HIGH;
    b.digitalWrite(greenLED, state);
    setTimeout(loop, main_delay);
    }
```

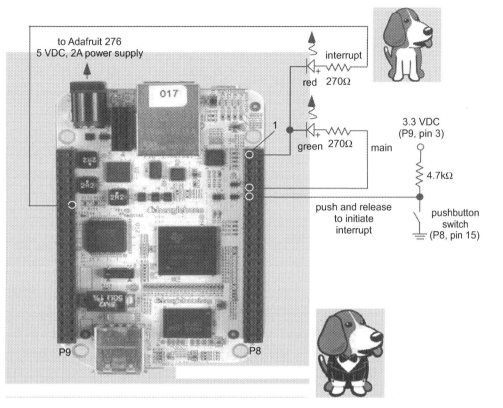

Figure 5.31: Bonescript interrupt example.In the main program loop a green LED is blinking at 100 ms intervals. When the switch is depressed and released, creating a rising edge on P8, pin 15 the ISR is executed which blinks a red LED at 50 ms intervals.

```
function blink_red()
{
b.digitalWrite(redLED, b.HIGH);
setTimeout(blink_red2, inter_delay);
}

function blink_red2()
{
b.digitalWrite(redLED, b.LOW);
setTimeout(blink_red3, inter_delay);
}
```

```
function blink_red3()
{
b.digitalWrite(redLED, b.HIGH);
setTimeout(blink_red4, inter_delay);
}

function blink_red4()
{
b.digitalWrite(redLED, b.LOW);
}
//*********************************************************
```

5.15   SUMMARY

In the first four chapters of this book we employed BeagleBone as a user–friendly processor. We accessed its features and subsystems via the Bonescript programming environment. In this chapter we began to shift focus and "unleash" the power of BeagleBone as a Linux–based, 32–bit, super–scalar ARM Cortex A8 processor. We began with a brief review of the C and C++ tool programming chain followed by examples on how to interact with the digital and analog pins aboard the processor. We then took a closer look at the features and subsystems available aboard BeagleBone. We spent a good part of the chapter describing the exposed functions of BeagleBone. These are the functions accessible to the user via the P8 and P9 extension headers. Throughout the chapter we provided sample programs on how to interact with and program the exposed functions.

5.16   REFERENCES

- Kelley, A. and I. Pohl. *A Book on C – Programming in C.* 4th edition. Boston, MA: Addison Wesley, 1998. Print.

- Vander Veer, Emily. *JavaScript for Dummies.* Hoboken, NJ: Wiley Publishing, Inc., 4th edition, 2005. Print.

- Pollock, John. *JavaScript.* New York, NY: McGraw Hill, 3rd edition, 2010. Print.

- Kiessling, Manuel. *The Node Beginner Guide: A Comprehensive Node.js Tutorial.* 2012. Print.

- Hughes—Croucher, Tom and Mike Wilson. *Node Up and Running.* Sebastopol, CA: O'Reilly Media, Inc., 2012. Print.

- von Hagen, W. *Ubuntu Linux Bible.* Indianapolis, IN: Wiley Publishing, Inc., 2007. Print.

- Coley, Gerald. *BeagleBone Rev A6 Systems Reference Manual.* Revision 0.0, May 9, 2012, beagleboard.org www.beaglebord.org

- Likely, Grant. *Device Tree*. `http://omappedia.org`.

- *Device Tree Usage*. `www.devicetree.org`

- *Enable PWM on BeagleBone with Device Tree Overlays*. `www.hipstercircuits.com`

- Korsch, James and Leonard Garrett. *Data Structures, Algorithms, and Program Style Using C*. Boston, MA: PWS–Kent Publishing Company, 1988. Print.

- Barrett, Steven and Daniel Pack. *Embedded Systems Design and Applications with the 68HC12 and HCS12*. Upper Saddle River, NJ: Pearson Prentice Hall, 2005. Print.

- Barrett, Steven and Daniel Pack. *Processors Fundamentals for Engineers and Scientists*. Morgan and Claypool Publishers, 2006. `www.morganclaypool.com`

- Barrett, Steven and Daniel Pack. *Atmel AVR Processor Primer Programming and Interfacing*. Morgan and Claypool Publishers, 2008. `www.morganclaypool.com`

- Horowitz, P. and W. Hill. *The Art of Electronics*. 2 nd edition, New York, NY: Cambridge University Press, 1990. Print.

- *AM335X PWMSS Driver's Guide*. Texas Instruments.

- *BeagleBone LCD7 Cape Rev A3 System Reference Manual*.

- *The I2C–Bus Specification*. Version 2.1, Philips Semiconductor, January 2000.

- *CAN–A Brief Tutorial for Embedded Engineers*, `www.computer-solutions.co.uk`

- SMSC. *LAN8710A/LAN8710Ai Small Footprint MII/RMII 10/100 Etthernet Transceiver with HP Auto –MDIX and flexPWR Technology*. August 12, 2012.

5.17 CHAPTER EXERCISES

1. What are the differences between Ubuntu and the Ångström Distribution release of Linux?

2. How does BeagleBone interact with a host computer?

3. Describe how to properly interface an LED to a processor.

4. Develop a glossary of Linux commands introduced in this chapter.

5. Given a sinusoid with 500 Hz frequency, what should be the minimum sampling frequency for an analog–to–digital converter, if we want to faithfully reconstruct the analog signal after the conversion?

6. If 12 bits are used to quantize a sampled signal, what is the number of available quantized levels? What will be the resolution of such a system if the input range of the analog–to–digital converter is 1.8 VDC?

7. A flex sensor provides 10K ohm of resistance for 0 degrees flexure and 40K ohm of resistance for 90 degrees of flexure. Design a circuit to convert the resistance change to a voltage change (Hint: consider a voltage divider). Then design a transducer interface circuit to convert the output from the flex sensor circuit to voltages suitable for the BeagleBone ADC system.

8. Does the time to convert an analog input signal to a digital representation vary in a successive-–approximation converter relative to the magnitude of the input signal? Explain.

9. Summarize the differences between the USART, SPI, and I2C methods of serial communication.

10. What is the primary difference between the UART and SPI serial communication systems?

11. What is the ASCII encoded value for "BeagleBone?"

12. What is the purpose of an interrupt?

13. Describe the flow of events when an interrupt occurs.

CHAPTER 6

BeagleBone "Off the Leash"

Objectives: After reading this chapter, the reader should be able to do the following:

- Enjoy the full power and rapid prototyping features of the Bonescript environment.

- Develop the hardware and Bonescript based control algorithm for a weather station.

- Develop the hardware and Bonescript based control algorithm for a Speak–and–Spell like device.

- Develop the hardware and a C based control algorithm for the Dagu Rover 5 treaded robot.

- Describe multiple Linux compatible open source libraries.

- Explore OpenCV computer vision features as a case study of available open source libraries.

- Program BeagleBone using OpenCV functions to capture and display facial images and place a moustache on the face.

- Construct Boneyard II – a portable BeagleBone platform.

6.1 OVERVIEW

In the early chapters of this book, we examined the Bonescript environment as a user–friendly tool to rapidly employ BeagleBone features right out of the box. In this chapter, we revisit Bonescript and demonstrate its power as a rapid prototyping tool to develop complex, processor–based, expandable systems employing multiple BeagleBone subsystems. Specifically we develop a Bonescript based weather station and Speak–and–Spell like device. We then illustrate C based system development. We construct a control system for a Dagu Rover 5 treaded robot. We use code developed in the previous chapters as building blocks for this system. We conclude the chapter by taking a brief look at the rich variety of open source libraries available to the Linux developer. As a case study, we review some of the fundamental features of the OpenCV computer vision library and employ them in a fun image processing task to plant moustaches on face images. As a Linux based processor with a clock speed as high as 1 GHz, BeagleBone is equipped to handle complex image processing tasks not possible with a microcontroller–based board.

     We have carefully chosen each of these projects to illustrate how BeagleBone may be used in a variety of project areas including instrumentation intensive applications (weather station), in assistive and educational technology applications (Speak–and–Spell), in motor and industrial control applications (Dagu robot), and calculation intensive image processing applications (moustache cam).

6.2   BONEYARD II: A PORTABLE LINUX PLATFORM–BEAGLEBONE UNLEASHED

In the first five chapters of the book BeagleBone was "leashed" to a host computer. This is a good way to efficiently use the features of the host computer in application development. However, BeagleBone can be quickly unleashed and converted into a standalone Linux computer as shown in Figure 6.1, 6.2 and 6.3. We have dubbed this project Boneyard II. This capability would be especially useful for developing applications that will be used in a remote, autonomous application.

Figure 6.1: Boneyard II–a standalone Linux computer (photo courtesy of J. Barrett, Closer to the Sun International).

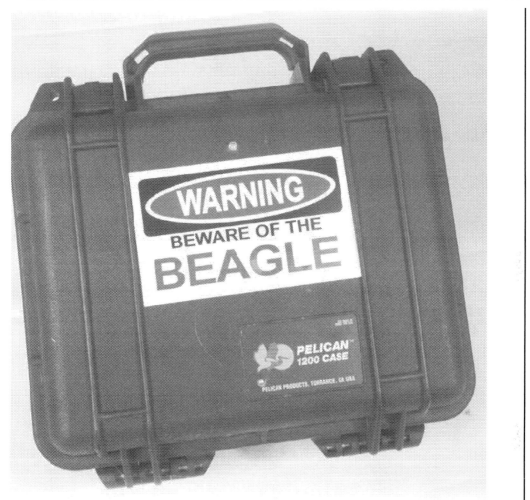

Figure 6.2: Boneyard II–a standalone Linux computer (photo courtesy of J. Barrett, Closer to the Sun International).

The Boneyard II is quickly assembled using off–the–shelf components including:

- an original BeagleBone (700 MHz) or black (1 GHz) processor board.

- the Circuitco LCD7 display

- a Adafruit mini–keyboard

- a USB hub

Figure 6.3: Boneyard II–a standalone Linux computer (photo courtesy of J. Barrett, Closer to the Sun International).

- a mini USB mouse

- and a Pelican 1200 case

All of the components may be purchased for under US$300. A configuration diagram is provided in 6.4. This is truly a plug–and–play system. Simply by connecting the components as shown and powering up the system, a standalone, 1 GHz computer may be assembled for under $300.

6.3   APPLICATION 1: WEATHER STATION IN BONESCRIPT

In the early chapters of this book, we examined the Bonescript environment as a user–friendly tool to rapidly employ BeagleBone features right out of the box. In this section, we demonstrate how Bonescript may be used as a rapid prototyping tool to develop complex, processor–based systems employing multiple BeagleBone subsystems. In this first example we develop a basic weather station to sense wind direction and ambient temperature. The weather station may be easily expanded with other sensors (wind speed, humidity, etc.) The sensed values will be displayed on an LCD in Fahrenheit. The wind direction will also be displayed on a multi–LED array.

6.3.1   REQUIREMENTS

The requirements for this system include:

- Design a weather station to sense wind direction and ambient temperature.

- Sensed wind direction and temperature will be displayed on an LCD.

- Sensed temperature will be displayed in the Fahrenheit temperature scale.

- Wind direction will be displayed on a multi–LED array.

6.3.2   STRUCTURE CHART

To begin the design process a structure chart is used to partition the system into definable pieces. We employ a top–down design/bottom–up implementation approach. The structure chart for the weather station is provided in Figure 6.5. The system is partitioned until the lowest level of the structure chart contains "doable" pieces of hardware components or software functions. Data flown is shown on the structure chart as directed arrows.

The main BeagleBone subsystem needed for this project is the ADC system to convert the analog voltage from the LM34 temperature sensor and weather vane into digital signals. Also, a number of general purpose input/output pins will interface to the wind direction display. The wind direction display consists of a multi–LED array. Each LED has a 2.1 VDC voltage drop and a current of 10 mA.

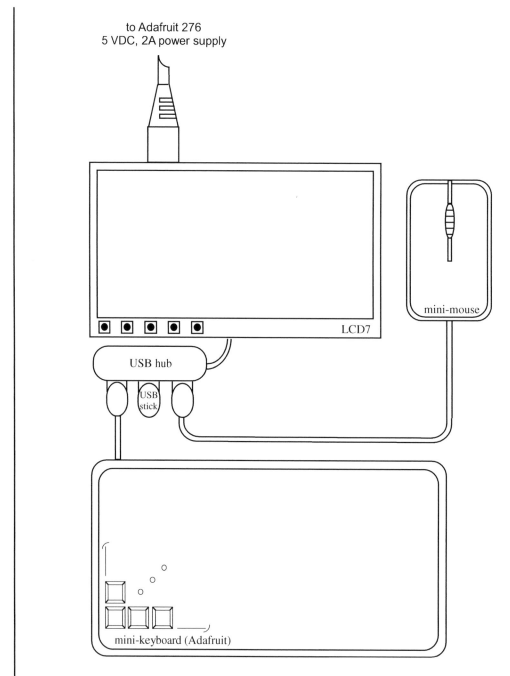

Figure 6.4: Boneyard II connection diagram.

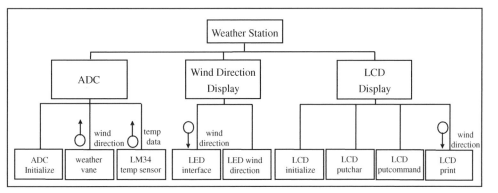

Figure 6.5: Weather station structure chart.

6.3.3 CIRCUIT DIAGRAM

The circuit diagram for the weather station is provided in Figure 6.6. The weather station is equipped with two input sensors: the LM34 to measure temperature and the weather vane to measure wind direction. Both of the sensors provide an analog output that is fed to BeagleBone's ADC system. The LM34 provides 10 mV output per degree Fahrenheit. The weather vane provides 0 to 1.8 VDC for 360 degrees of vane rotation. The weather vane must be oriented to a known direction with the output voltage at this direction noted. We assume that 0 VDC corresponds to North and the voltage increases as the vane rotates clockwise to the East. The vane output voltage continues to increase until North is again reached at 1.8 VDC and then rolls over back to 0 volts. All other directions are derived from this reference point as shown in Figure 6.7. An LCD is connected to BeagleBone as shown in Figure 6.6. This is the same LCD interface provided earlier in the book.

6.3.4 UML ACTIVITY DIAGRAMS

The UML activity diagram for the program is provided in Figure 6.8. After initializing the subsystems, the program enters a continuous loop where temperature and wind direction are sensed and displayed on the LCD and the LED display. The system then enters a delay. The delay value is set to determine how often the temperature and wind direction parameters are updated.

6.3.5 BONESCRIPT CODE

In this example we use the Bonescript user environment to rapidly code the control algorithm for the weather station. We use examples provided earlier in the book as building blocks to rapidly construct the code. We provide the majority of the code. The code to convert the reading from the LM34 temperature sensor and display its value and wind direction are left as an end of chapter exercise.

//***

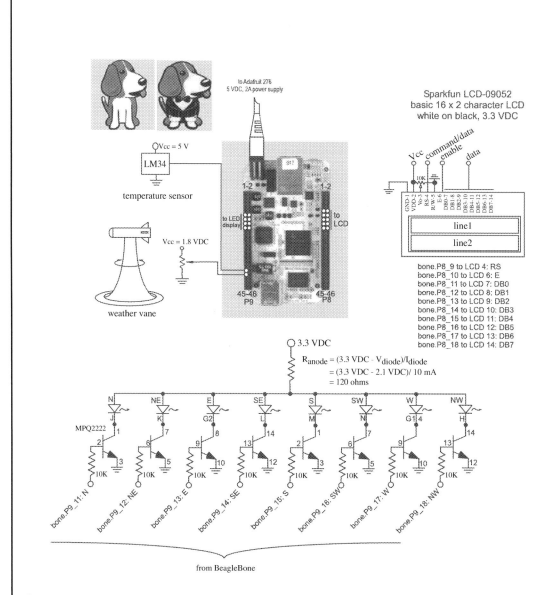

Figure 6.6: Circuit diagram for weather station.

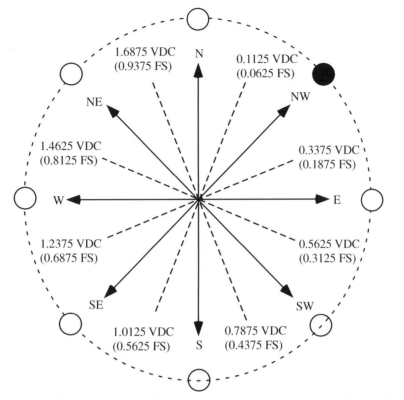

Figure 6.7: Weather vane output voltage as shown as the actual value and the normalized full scale (FS) value.

```
var b = require ('bonescript');

//Old bonescript defines 'bone' globally
var pins = (typeof bone != 'undefined') ? bone : b.bone.pins;

//sensor pin configuration
var wx_vane = pins.P9_39;              //weather vane
var temp_sen= pins.P9_37;              //temperature sensor
var wx_vane_value;
var temp_sen_value;

//wind direction LED
var LED_N   = pins.P9_11;              //N:  LED segment J
```

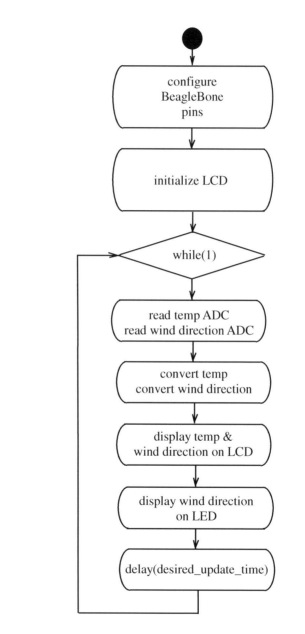

Figure 6.8: Weather station UML activity diagram.

```
var LED_NE  = pins.P9_12;          //NE: LED segment K
var LED_E   = pins.P9_13;          //E:  LED segment G2
var LED_SE  = pins.P9_14;          //SE: LED segment L
var LED_S   = pins.P9_15;          //S:  LED segment M
var LED_SW  = pins.P9_16;          //SW: LED segment N
var LED_W   = pins.P9_17;          //W:  LED segment G1
var LED_NW  = pins.P9_18;          //NW: LED segment H


//LCD pin configuration
var LCD_RS  = pins.P8_9;           //LCD Register Set (RS) control
var LCD_E   = pins.P8_10;          //LCD Enable (E) control
var LCD_DB0 = pins.P8_11;          //LCD Data line DB0
var LCD_DB1 = pins.P8_12;          //LCD Data line DB1
var LCD_DB2 = pins.P8_13;          //LCD Data line DB2
var LCD_DB3 = pins.P8_14;          //LCD Data line DB3
var LCD_DB4 = pins.P8_15;          //LCD Data line DB4
var LCD_DB5 = pins.P8_16;          //LCD Data line DB5
var LCD_DB6 = pins.P8_17;          //LCD Data line DB6
var LCD_DB7 = pins.P8_18;          //LCD Data line DB7


//wind direction pins
b.pinMode(LED_N,  b.OUTPUT);
b.pinMode(LED_NE, b.OUTPUT);
b.pinMode(LED_E,  b.OUTPUT);
b.pinMode(LED_SE, b.OUTPUT);
b.pinMode(LED_S,  b.OUTPUT);
b.pinMode(LED_SW, b.OUTPUT);
b.pinMode(LED_W,  b.OUTPUT);
b.pinMode(LED_NW, b.OUTPUT);


//LCD direction pins
b.pinMode(LCD_RS,  b.OUTPUT);      //set pin to digital output
b.pinMode(LCD_E,   b.OUTPUT);      //set pin to digital output
b.pinMode(LCD_DB0, b.OUTPUT);      //set pin to digital output
b.pinMode(LCD_DB1, b.OUTPUT);      //set pin to digital output
b.pinMode(LCD_DB2, b.OUTPUT);      //set pin to digital output
b.pinMode(LCD_DB3, b.OUTPUT);      //set pin to digital output
b.pinMode(LCD_DB4, b.OUTPUT);      //set pin to digital output
b.pinMode(LCD_DB5, b.OUTPUT);      //set pin to digital output
```

```
b.pinMode(LCD_DB6, b.OUTPUT);            //set pin to digital output
b.pinMode(LCD_DB7, b.OUTPUT);            //set pin to digital output
LCD_init();                             //call LCD initialize

setInterval(updateWeather, 100);

function updateWeather()
  {
  clear_LEDs();

  //Read weather vane and temperature sensors
  wx_vane_value  = b.analogRead(wx_vane);
  temp_sen_value = b.analogRead(temp_sen);

  //Calculate temperature

  //North
  if((wx_vane_value > 0.9375)||(wx_vane_value <= 0.0625))
    {
    //illuminate N LED
    b.digitalWrite(LED_N, b.HIGH);
    }

 //Northeast
 else if((wx_vane_value > 0.0625)&&(wx_vane_value <= 0.1875))
   {
   //illuminate NE LED
   b.digitalWrite(LED_NE, b.HIGH);
   }

 //East
 else if((wx_vane_value > 0.1875)&&(wx_vane_value <= 0.3125))
   {
   //illuminate E LED
   b.digitalWrite(LED_E, b.HIGH);
   }

 //Southeast
 else if((wx_vane_value > 0.3125)&&(wx_vane_value <= 0.4375))
```

```
    {
    //illuminate SE LED
    b.digitalWrite(LED_SE, b.HIGH);
    }

  //South
  else if((wx_vane_value > 0.4375)&&(wx_vane_value <= 0.5625))
    {
    //illuminate S LED
    b.digitalWrite(LED_S, b.HIGH);
    }

  //Southwest
  else if((wx_vane_value > 0.5625)&&(wx_vane_value <= 0.6875))
    {
    //illuminate SW LED
    b.digitalWrite(LED_SW, b.HIGH);
    }

  //West
  else if((wx_vane_value > 0.6875)&&(wx_vane_value <= 0.8125))
    {
    //illuminate W LED
    b.digitalWrite(LED_W, b.HIGH)
    }

  //NE
  else
    {
    //illuminate NE LED
    b.digitalWrite(LED_NE, b.HIGH)
    }
   }

//****************************************************************
//clear_LEDs
//****************************************************************

function clear_LEDs() {
```

```
   //reset LEDs
   b.digitalWrite(LED_N,  b.LOW);
   b.digitalWrite(LED_NE, b.LOW);
   b.digitalWrite(LED_E,  b.LOW);
   b.digitalWrite(LED_SE, b.LOW);
   b.digitalWrite(LED_S,  b.LOW);
   b.digitalWrite(LED_SW, b.LOW);
   b.digitalWrite(LED_W,  b.LOW);
   b.digitalWrite(LED_NW, b.LOW);
   }

//******************************************************************
//LCD_print
//******************************************************************

function LCD_print(line, message, callback)
{
var i = 0;

if(line == 1)
  {
  LCD_putcommand(0x80, writeNextCharacter);//print to LCD line 1
  }
else
  {
  LCD_putcommand(0xc0, writeNextCharacter);//print to LCD line 2
  }

function writeNextCharacter()
  {
  //if we already printed the last character, stop and callback
  if(i == message.length)
    {
    if(callback) callback();
    return;
    }

  //get the next character to print
  var chr = message.substring(i, i+1);
```

```
  i++;

  //print it using LCD_putchar and come back again when done
  LCD_putchar(chr, writeNextCharacter);
  }
}


//****************************************************************
//LCD_init
//****************************************************************

function LCD_init(callback)
{
//LCD Enable (E) pin low
b.digitalWrite(LCD_E, b.LOW);

//Start at the beginning of the list of steps to perform
var i = 0;

//List of steps to perform
var steps =
  [
  function(){ setTimeout(next, 15); },         //delay 15ms
  function(){ LCD_putcommand(0x38, next); },   //set for 8-bit operation
  function(){ setTimeout(next, 5); },          //delay 5ms
  function(){ LCD_putcommand(0x38, next); },   //set for 8-bit operation
  function(){ LCD_putcommand(0x38, next); },   //set for 5 x 7 character
  function(){ LCD_putcommand(0x0E, next); },   //display on
  function(){ LCD_putcommand(0x01, next); },   //display clear
  function(){ LCD_putcommand(0x06, next); },   //entry mode set
  function(){ LCD_putcommand(0x00, next); },   //clear display, cursor home
  function(){ LCD_putcommand(0x00, callback); } //clear display, cursor home
  ];

next(); //Execute the first step

//Function for executing the next step
function next()
  {
```

```
  i++;
  steps[i-1]();
  }
}


//**********************************************************************
//LCD_putcommand
//**********************************************************************

function LCD_putcommand(cmd, callback)
{
//parse command variable into individual bits for output
//to LCD
if((cmd & 0x0080)== 0x0080) b.digitalWrite(LCD_DB7, b.HIGH);
  else b.digitalWrite(LCD_DB7, b.LOW);
if((cmd & 0x0040)== 0x0040) b.digitalWrite(LCD_DB6, b.HIGH);
  else b.digitalWrite(LCD_DB6, b.LOW);
if((cmd & 0x0020)== 0x0020) b.digitalWrite(LCD_DB5, b.HIGH);
  else b.digitalWrite(LCD_DB5, b.LOW);
if((cmd & 0x0010)== 0x0010) b.digitalWrite(LCD_DB4, b.HIGH);
  else b.digitalWrite(LCD_DB4, b.LOW);
if((cmd & 0x0008)== 0x0008) b.digitalWrite(LCD_DB3, b.HIGH);
  else b.digitalWrite(LCD_DB3, b.LOW);
if((cmd & 0x0004)== 0x0004) b.digitalWrite(LCD_DB2, b.HIGH);
  else b.digitalWrite(LCD_DB2, b.LOW);
if((cmd & 0x0002)== 0x0002) b.digitalWrite(LCD_DB1, b.HIGH);
  else b.digitalWrite(LCD_DB1, b.LOW);
if((cmd & 0x0001)== 0x0001) b.digitalWrite(LCD_DB0, b.HIGH);
  else b.digitalWrite(LCD_DB0, b.LOW);

//LCD Register Set (RS) to logic zero for command input
b.digitalWrite(LCD_RS, b.LOW);
//LCD Enable (E) pin high
b.digitalWrite(LCD_E, b.HIGH);

//End the write after 1ms
setTimeout(endWrite, 1);

function endWrite()
```

```
  {
  //LCD Enable (E) pin low
  b.digitalWrite(LCD_E, b.LOW);
  //delay 1ms before calling 'callback'
  setTimeout(callback, 1);
  }
}


//*******************************************************************
//LCD_putchar
//*******************************************************************

function LCD_putchar(chr1, callback)
{
//Convert chr1 variable to UNICODE (ASCII)
var chr = chr1.toString().charCodeAt(0);

//parse character variable into individual bits for output
//to LCD
if((chr & 0x0080)== 0x0080) b.digitalWrite(LCD_DB7, b.HIGH);
  else b.digitalWrite(LCD_DB7, b.LOW);
if((chr & 0x0040)== 0x0040) b.digitalWrite(LCD_DB6, b.HIGH);
  else b.digitalWrite(LCD_DB6, b.LOW);
if((chr & 0x0020)== 0x0020) b.digitalWrite(LCD_DB5, b.HIGH);
  else b.digitalWrite(LCD_DB5, b.LOW);
if((chr & 0x0010)== 0x0010) b.digitalWrite(LCD_DB4, b.HIGH);
  else b.digitalWrite(LCD_DB4, b.LOW);
if((chr & 0x0008)== 0x0008) b.digitalWrite(LCD_DB3, b.HIGH);
  else b.digitalWrite(LCD_DB3, b.LOW);
if((chr & 0x0004)== 0x0004) b.digitalWrite(LCD_DB2, b.HIGH);
  else b.digitalWrite(LCD_DB2, b.LOW);
if((chr & 0x0002)== 0x0002) b.digitalWrite(LCD_DB1, b.HIGH);
  else b.digitalWrite(LCD_DB1, b.LOW);
if((chr & 0x0001)== 0x0001) b.digitalWrite(LCD_DB0, b.HIGH);
  else b.digitalWrite(LCD_DB0, b.LOW);

//LCD Register Set (RS) to logic one for character input
b.digitalWrite(LCD_RS, b.HIGH);
//LCD Enable (E) pin high
```

```
b.digitalWrite(LCD_E, b.HIGH);

//End the write after 1ms
setTimeout(endWrite, 1);

function endWrite()
  {
  //LCD Enable (E) pin low and call scheduleCallback when done
  b.digitalWrite(LCD_E, b.LOW);
  //delay 1ms before calling 'callback'
  setTimeout(callback, 1);
  }
}

//*****************************************************************
```

6.4 APPLICATION 2: SPEAK-AND-SPELL IN C

Speak–and–Spell is an educational toy developed by Texas Instruments in the mid–1970's. It was developed by the engineering team of Gene Franz, Richard Wiggins, Paul Breedlove and Larry Branntingham pictured in Figure 6.9. The Speak–and–Spell consists of a keyboard, display, speech synthesizer, and a slot to insert game modules. A series of educational games teach spelling skills, letter recognition skills, and memory aids are available as plug in cartridges [www.ti.com].

In this project we design a BeagleBone based Speak–and–Spell. We use a small keyboard (www.adafruit.com) connected to BeagleBone via the USB port. We also use the Circuitco 7 inch LCD display (BeagleBone LCD7). For speech synthesis we use the SP0–512 text to speech chip (www.speechchips.com). The SP0–512 accepts UART compatible serial text stream. The text stream is converted to phoneme codes used to generate an audio output. The chip requires a 9600 Baud bit stream with no parity, 8 data bits and a stop bit. Additional information on the chip and its features are available at www.speechchips.com. The BeagleBone version of Speak–and–Spell is shown in Figure 6.10 and the support circuit for the SP0–512 is provided in Figure 6.11.

6.4.1 BEAGLEBONE C CODE

The structure chart and UML activity diagram for Speak–and–Spell is provided in Figure 6.12. A basic algorithm is provided below to accept input from the keyboard, output it on the LCD7 display and pass it to the SP0–512 speech synthesizer chip. This algorithm may form the basis for a number of Speak–and–Spell educational games.

Original BeagleBone–Linux 3.2

```
//*****************************************************************
```

Figure 6.9: Speak–and–Spell design team from left to right: Gene Franz, Richard Wiggins, Paul Breedlove and Larry Branntingham [`www.ti.com`].

```
//sns.cpp - Speak-and-Spell
//for BeagleBone original with Linux 3.2
//  - prompts user for input
//  - prints input to screen
//  - configures BeagleBone P8, pin 5 for receive from SPO-512
//    speech chip SPEAKING pin (SPO-512, pin 17)
//  - configures BeagleBone uart1 for transmission and 9600 Baud
//    (P9, 24)
//  - provides spoken input via speech synthesis chip connected
//    to uart1 tx pin (P9, 24)
//*********************************************************************

#include <stdio.h>
#include <stdlib.h>
```

Figure 6.10: BeagleBone based Speak–and–Spell (photo courtesy of J. Barrett, Closer to the Sun International).

```
#include <stddef.h>
#include <time.h>
#include <iostream>
#include <termios.h>
#include <fcntl.h>
#include <unistd.h>
#include <sys/types.h>
#include <string.h>
#include <sys/time.h>

#define  output  "out"
#define  input   "in"
```

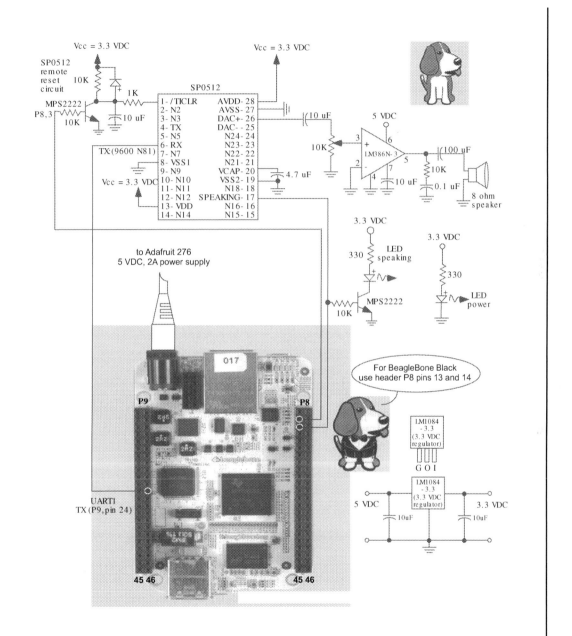

Figure 6.11: Speech synthesis support circuit [www.speechchips.com]. **Note:** For BeagleBone black the SP0512 remote reset circuit should be connected to header P8 pin 13.

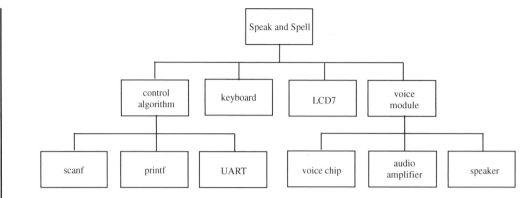

Figure 6.12: Speak–and–Spell structure chart.

```
//function prototype
void delay_us(int);
void delay_sec(float delay_value);
void configure_uart1(void);
void configure_SP0_512(void);
void SP0_512_handshake(void);
void reset_SP0_512(void);
void configure_P8_3_output(void);

//global variables
//***UART1***
//define file handle for uart1
FILE *ofp_uart1_tx, *ofp_uart1_rx;

//define mux mode for uart1 tx
int mux_mode_uart1_tx = 0X8000;
int mux_mode_uart1_rx = 0X8020;

//uart1 configuration using termios
termios uart1;
int fd, i;
char byte_out[40];

//***SP0-512***
```

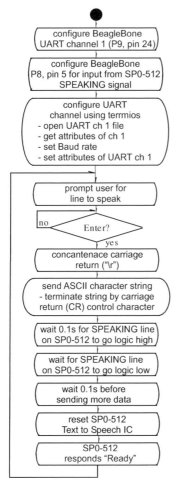

Figure 6.13: Speak–and–Spell UML activity diagram.

```
//configure P8, pin 5 for input - connect to SP0-512 SPEAKING signal
//define file handles for gpio34 (P8, pin 5, GPIO1_2)
FILE *ofp_gpmc_ad2, *ofp_export_34, *ifp_gpio34_value, *ofp_gpio34_direction;

//define pin variables for gpio34
int mux_mode_34 = 0x003f, pin_number_34 = 34, logic_status_34;
char* pin_direction_34 = input;

//configure P8, pin 3 for output - connect to reset circuit on SP0-512
```

```
//define file handles for gpio38 (P8, pin 3, GPIO1_6)
FILE *ofp_gpmc_ad6, *ofp_export_38, *ofp_gpio38_value, *ofp_gpio38_direction;

//define pin variables for gpio38
int mux_mode_38 = 0x0007, pin_number_38 = 38, logic_status_38 = 0;
char* pin_direction_38 = output;

char cntr_cr[] = "\r";
char reset[] = "[RESET]";

int main(void)
{
configure_uart1();
configure_P8_3_output();
configure_SP0_512();

while(1)
  {
  printf("Enter letter, word, statement.
Press [Enter].\r\n\n");
  scanf("%s", byte_out);
  printf("%s\r\n\n", byte_out);
  strcat(byte_out, cntr_cr);
  write(fd, byte_out, strlen(byte_out)+2);
  SP0_512_handshake();
  reset_SP0_512();
  }

fclose(ofp_gpmc_ad2);
fclose(ofp_export_34);
fclose(ofp_gpio34_direction);
fclose(ifp_gpio34_value);
close(fd);
}

//****************************************************************

void SP0_512_handshake(void)
{
```

```c
  for(i=0; i<=10; i++)                //wait for SPEAKING line to go logic high
    {
    delay_us(10000);
    }

  logic_status_34 = 1;

  while(logic_status_34 != 0)    //wait for SPEAKING line to go low
    {
    //configure gpio34 value and read the gpio34 pin
    ifp_gpio34_value = fopen("/sys/class/gpio/gpio34/value", "r");
    if(ifp_gpio34_value == NULL) {printf("Unable to open gpio34_value.\n");}
    fseek(ifp_gpio34_value, 0, SEEK_SET);
    fscanf(ifp_gpio34_value, "%d", &logic_status_34);
    fclose(ifp_gpio34_value);
    }

  for(i=0; i<=10; i++)                //wait for SPEAKING line to go logic high
    {
    delay_us(10000);
    }
}


//***********************************************************

void configure_uart1(void)
{

//configure uart1 for transmission
ofp_uart1_tx = fopen("/sys/kernel/debug/omap_mux/uart1_txd", "w");
if(ofp_uart1_tx == NULL) printf("Unable to open uart1 tx file.\n");
fseek(ofp_uart1_tx, 0, SEEK_SET);
fprintf(ofp_uart1_tx, "0x%2x", mux_mode_uart1_tx);
fflush(ofp_uart1_tx);

//configure uart1 for reception
ofp_uart1_rx = fopen("/sys/kernel/debug/omap_mux/uart1_rxd", "w");
if(ofp_uart1_rx == NULL) printf("Unable to open uart1 rx file.\n");
fseek(ofp_uart1_rx, 0, SEEK_SET);
```

```
    fprintf(ofp_uart1_rx, "0x%2x", mux_mode_uart1_rx);

    //open uart1 for tx/rx, not controlling device
    if((fd = open("/dev/ttyO1", O_RDWR | O_NOCTTY)) < 0)
      printf("Unable to open uart1 access.\n");

    //get attributes of uart1
    if(tcgetattr(fd, &uart1) < 0)
      printf("Could not get attributes of UART1 at ttyO1\n");

    //set Baud rate
    if(cfsetospeed(&uart1, B9600) < 0)
      printf("Could not set baud rate\n");
    else
      printf("Baud rate: 9600\n");

    //set attributes of uart1
    uart1.c_iflag = 0;
    uart1.c_oflag = 0;
    uart1.c_lflag = 0;
    uart1.c_cflag |= CS8;
    tcsetattr(fd, TCSANOW, &uart1);
    }


//****************************************************************

void configure_SP0_512(void)
{
//gpio34 mux setting
ofp_gpmc_ad2 = fopen("/sys/kernel/debug/omap_mux/gpmc_ad2", "w");
if(ofp_gpmc_ad2 == NULL) {printf("Unable to open gpmc_ad2.\n");}
fseek(ofp_gpmc_ad2, 0, SEEK_SET);
fprintf(ofp_gpmc_ad2, "0x%02x", mux_mode_34);
fflush(ofp_gpmc_ad2);

//create direction and value file for gpio34
ofp_export_34 = fopen("/sys/class/gpio/export", "w");
if(ofp_export_34 == NULL) {printf("Unable to open export.\n");}
fseek(ofp_export_34, 0, SEEK_SET);
```

```
fprintf(ofp_export_34, "%d", pin_number_34);
fflush(ofp_export_34);


//configure gpio34 direction
ofp_gpio34_direction = fopen("/sys/class/gpio/gpio34/direction", "w");
if(ofp_gpio34_direction == NULL) {printf("Unable to open gpio34_direction.\n");}
fseek(ofp_gpio34_direction, 0, SEEK_SET);
fprintf(ofp_gpio34_direction, "%s",  pin_direction_34);
}


//**************************************************************************

void configure_P8_3_output(void)
{
//gpio38 mux setting
ofp_gpmc_ad6 = fopen("/sys/kernel/debug/omap_mux/gpmc_ad6", "w");
if(ofp_gpmc_ad6 == NULL) {printf("Unable to open gpmc_ad6.\n");}
fseek(ofp_gpmc_ad6, 0, SEEK_SET);
fprintf(ofp_gpmc_ad6, "0x%02x", mux_mode_38);
fflush(ofp_gpmc_ad6);


//create direction and value file for gpio38
ofp_export_38 = fopen("/sys/class/gpio/export", "w");
if(ofp_export_38 == NULL) {printf("Unable to open export.\n");}
fseek(ofp_export_38, 0, SEEK_SET);
fprintf(ofp_export_38, "%d", pin_number_38);
fflush(ofp_export_38);


//configure gpio38 direction
ofp_gpio38_direction = fopen("/sys/class/gpio/gpio38/direction", "w");
if(ofp_gpio38_direction == NULL) {printf("Unable to open gpio38_direction.\n");}
fseek(ofp_gpio38_direction, 0, SEEK_SET);
fprintf(ofp_gpio38_direction, "%s",  pin_direction_38);
fflush(ofp_gpio38_direction);


//configure gpio38 value - initially set logic low
ofp_gpio38_value = fopen("/sys/class/gpio/gpio38/value", "w");
if(ofp_gpio38_value == NULL) {printf("Unable to open gpio38_value.\n");}
fseek(ofp_gpio38_value, 0, SEEK_SET);
```

```
fprintf(ofp_gpio38_value, "%d", logic_status_38);
fflush(ofp_gpio38_value);
}

//*************************************************************

void reset_SP0_512(void)
{
logic_status_38 = 1;
fseek(ofp_gpio38_value, 0, SEEK_SET);
fprintf(ofp_gpio38_value, "%d", logic_status_38);
fflush(ofp_gpio38_value);

delay_us(10000);

logic_status_38 = 0;
fseek(ofp_gpio38_value, 0, SEEK_SET);
fprintf(ofp_gpio38_value, "%d", logic_status_38);
fflush(ofp_gpio38_value);
}

//*************************************************************

void delay_us(int desired_delay_us)
{
struct timeval  tv_start;  //start time hack
struct timeval  tv_now;    //current time hack
int elapsed_time_us;

gettimeofday(&tv_start, NULL);
elapsed_time_us = 0;

while(elapsed_time_us <  desired_delay_us)
  {
  gettimeofday(&tv_now, NULL);
  if(tv_now.tv_usec >= tv_start.tv_usec)
    elapsed_time_us = tv_now.tv_usec - tv_start.tv_usec;
  else
    elapsed_time_us = (1000000 - tv_start.tv_usec) + tv_now.tv_usec;
```

```c
    //printf("start: %ld \n", tv_start.tv_usec);
    //printf("now: %ld \n", tv_now.tv_usec);
    //printf("desired: %d \n", desired_delay_ms);
    //printf("elapsed: %d \n\n", elapsed_time_ms);
    }
}


//****************************************************************
//delay_sec(float delay_value)
//****************************************************************

void delay_sec(float delay_value)
{
time_t   now, later;

now = time(NULL);
later = time(NULL);

while(difftime(later,now) < delay_value)
  {
  later = time(NULL);
  }
}


//****************************************************************
      BeagleBone Black–Linux 3.8
//*******************************************************************
//sns.cpp - Speak and Spell
//  - prompts user for input
//  - prints input to screen
//  - provides spoken input via speech synthesis chip connected
//    to uart1
//  - configures BeagleBone uart1 for transmission and 9600 Baud
//
//Requires use of the 'uart1.dts' from Chapter 5
//*******************************************************************

#include <stdio.h>
#include <stdlib.h>
```

```cpp
#include <stddef.h>
#include <time.h>
#include <iostream>
#include <termios.h>
#include <fcntl.h>
#include <unistd.h>
#include <sys/types.h>
#include <string.h>

int main(void)
{
//define file handle for uart1
FILE *ofp_uart1_tx, *ofp_uart1_rx;

//uart1 configuration using termios
termios uart1;
int fd;

//open uart1 for tx/rx, not controlling device
if((fd = open("/dev/tty01", O_RDWR | O_NOCTTY)) < 0)
  printf("Unable to open uart1 access.\n");

//get attributes of uart1
if(tcgetattr(fd, &uart1) < 0)
  printf("Could not get attributes of UART1 at tty01\n");

//set Baud rate
if(cfsetospeed(&uart1, B9600) < 0)
  printf("Could not set baud rate\n");
else
  printf("Baud rate: 9600\n");

//set attributes of uart1
uart1.c_iflag = 0;
uart1.c_oflag = 0;
uart1.c_lflag = 0;
tcsetattr(fd, TCSANOW, &uart1);

char byte_out[20];
```

```
//set ASCII character G repeatedly
while(1)
  {
  printf("Enter letter, word, statement.
Press [Enter].\n\n");
  scanf("%s", byte_out);
  printf("%s\n\n\n", byte_out);
  write(fd, byte_out, strlen(byte_out)+1);
  }

close(fd);
}
//*********************************************************************
```

6.5   APPLICATION 3: DAGU ROVER 5 TREADED ROBOT

In this example we control a Dagu Rover ROV5–1 robot with a C based control system hosted on BeagleBone. The goal of the robot system is to navigate through the three–dimensional mountain maze described earlier in the book.

6.5.1   DESCRIPTION

Dagu manufactures a number of low cost educational robots and robotic arms. In this example we use the Dagu Rover ROV5–1 robot chassis. This robot is equipped with two motor driven treads. Dagu offers other robot configurations with additional motors and wheel encoders. We begin by equipping the ROV 5–1 with a plexi–glass platform, three IR sensors and a motor control interface. The robot platform is illustrated in Figure 6.14 and 6.15.

6.5.2   REQUIREMENTS

The requirements for this project are simple, the robot must autonomously navigate through the three–dimensional mountain maze without touching maze walls.

6.5.3   CIRCUIT DIAGRAM

The circuit diagram for the robot is provided in Figure 6.16. The three IR sensors (left, middle, and right) are mounted on the leading edge of the robot to detect maze walls. The interface for the IR sensors was used earlier in the book on other projects. The sensor outputs are fed to three ADC channels (AIN0, AIN1 and AIN2). The robot motors are driven by PWM channels EHRPWM1A (P9, pin 14) and B (P9, pin 16). BeagleBone is interfaced to the motors via a Darlington transistor

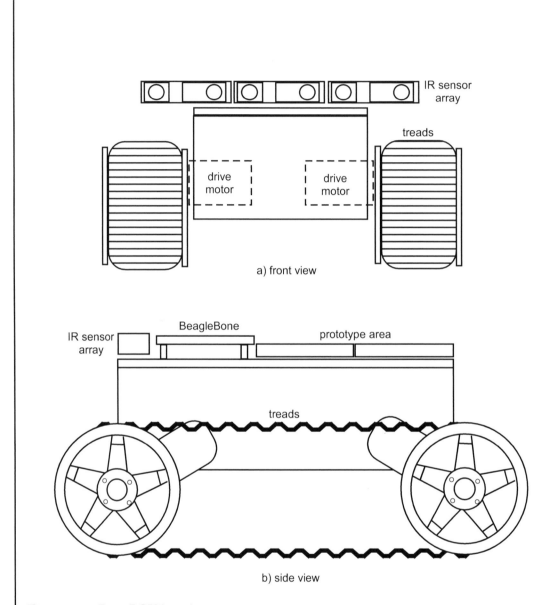

Figure 6.14: Dagu ROV5–1 robot.

Figure 6.15: Dagu ROV5–1 robot in 3D maze (photo courtesy of J. Barrett, Closer to the Sun International).

(TIP 120) with enough drive capability to handle the maximum current requirements of the motor. A 330 ohm resistor is used to limit base current to 5.5 mA. The resulting collector current and hence motor drive current is approximately 300 mA. The robot is powered via an external 7.2 VDC power supply umbilical to conserve battery use. The 7.2 VDC supply is routed through the 5 VDC and 3.3 VDC regulator matrix as shown in Figure 6.16.

6.5.4   STRUCTURE CHART

The structure chart for the robot project is provided in Figure 6.17. The two main systems used in this project is the PWM system to drive the motorized treads and the ADC system to read the IR sensors.

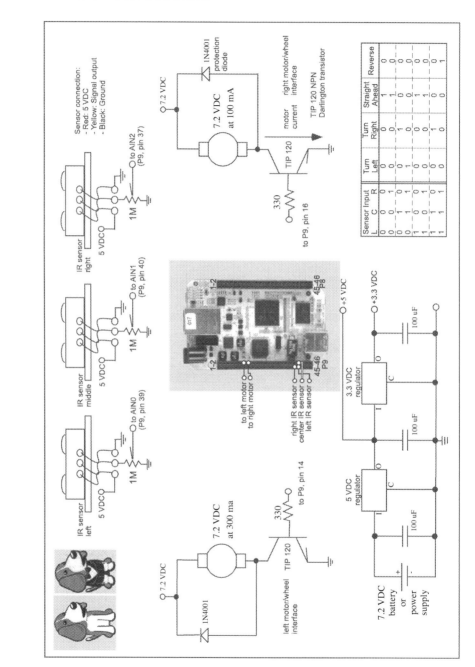

Figure 6.16: Dagu robot circuit diagram.

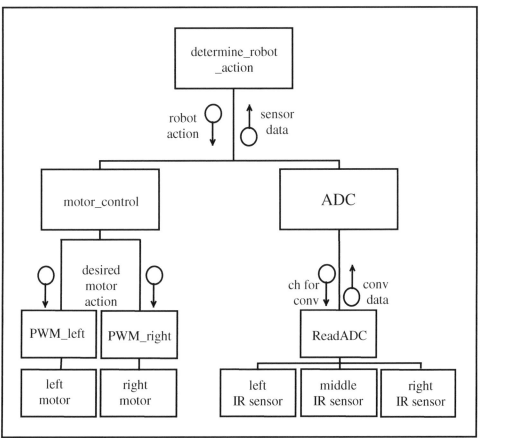

Figure 6.17: Dagu robot structure diagram.

6.5.5   UML ACTIVITY DIAGRAMS

The UML activity diagram for the robot is provided in Figure 6.18. The basic algorithm is quite straight forward. The sensor values are read and PWM command signals are issued to navigate about the maze.

6.5.6   BEAGLEBONE C CODE

As before we use code developed in the previous chapter as building blocks to rapidly develop the control algorithm for the Dagu robot. The printf statements are useful for algorithm development and troubleshooting. They should be commented out before testing the robot in the maze. **Original BeagleBone–Linux 3.2**

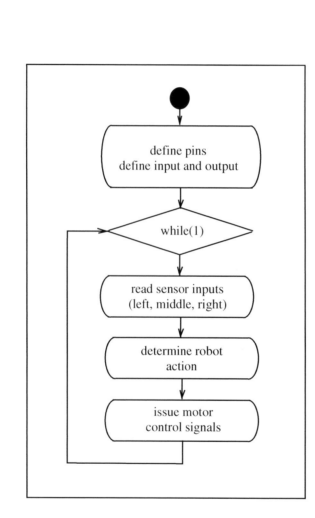

Figure 6.18: Dagu robot UML activity diagram.

```c
//*******************************************************************
//for BeagleBone original with Linux 3.2
//*******************************************************************

#include <stdio.h>
#include <stddef.h>
#include <time.h>
#include <math.h>

#define  output "out"
#define  input  "in"

int  main (void)
{
//wall detection threshold
double threshold = 2500.0;                  //experimentally determined

//configure adc channels
//define file handles for adc related files
FILE *ifp_ain1, *ifp_ain2, *ifp_ain3;
float ain0_value, ain1_value, ain2_value;

//open adc related files for access to ain0, 1 and 2
ifp_ain1 = fopen("/sys/bus/platform/devices/tsc/ain1", "r");
if (ifp_ain1 == NULL) {printf("Unable to ain1.\n");}

ifp_ain2 = fopen("/sys/bus/platform/devices/tsc/ain2", "r");
if (ifp_ain2 == NULL) {printf("Unable to ain2.\n");}

ifp_ain3 = fopen("/sys/bus/platform/devices/tsc/ain3", "r");
if (ifp_ain3 == NULL) {printf("Unable to ain3.\n");}

//configure pwm channels 0 and 1
//define file handles for channel 0 - EHRPWM1A (P9, pin 14)
//designated as ehrpwm.1:0
FILE *ofp_gpmc_a2, *pwm_freq0, *pwm_req0, *pwm_duty0;
FILE *pwm_polarity0, *pwm_run0;

//define pin variables for channel 0
```

```
int mux_mode0 = 0x006, request0 = 1, freq0 = 100, duty0 = 50;
int polarity0 = 1, run0 = 1;

ofp_gpmc_a2 = fopen("/sys/kernel/debug/omap_mux/gpmc_a2", "w");
if(ofp_gpmc_a2 == NULL) {printf("Unable to open gpmc_a2.\n");}
fseek(ofp_gpmc_a2, 0, SEEK_SET);
fprintf(ofp_gpmc_a2, "0x%02x", mux_mode0);
fflush(ofp_gpmc_a2);

pwm_req0 = fopen("/sys/class/pwm/ehrpwm.1:0/request", "w");
if(pwm_req0 == NULL) {printf("Unable to open pwm 0 request.\n");}
fseek(pwm_req0, 0, SEEK_SET);
fprintf(pwm_req0, "%d", request0);
fflush(pwm_req0);

pwm_freq0 = fopen("/sys/class/pwm/ehrpwm.1:0/period_freq", "w");
if(pwm_freq0 == NULL) {printf("Unable to open pwm 0 frequency.\n");}
fseek(pwm_freq0, 0, SEEK_SET);
fprintf(pwm_freq0, "%d", freq0);
fflush(pwm_freq0);

pwm_duty0 = fopen("/sys/class/pwm/ehrpwm.1:0/duty_percent", "w");
if(pwm_duty0 == NULL) {printf("Unable to open pwm 0 duty cycle.\n");}
fseek(pwm_duty0, 0, SEEK_SET);
fprintf(pwm_duty0, "%d", duty0);
fflush(pwm_duty0);

pwm_polarity0 = fopen("/sys/class/pwm/ehrpwm.1:0/polarity", "w");
if(pwm_polarity0 == NULL) {printf("Unable to open pwm 0 polarity.\n");}
fseek(pwm_polarity0, 0, SEEK_SET);
fprintf(pwm_polarity0, "%d", polarity0);
fflush(pwm_polarity0);

pwm_run0 = fopen("/sys/class/pwm/ehrpwm.1:0/run", "w");
if(pwm_run0 == NULL) {printf("Unable to open pwm 0 run.\n");}

//define file handles for channel 1 - EHRPWM1B (P9, pin 16)
//designated as ehrpwm.1:1
FILE *ofp_gpmc_a3, *pwm_freq1, *pwm_req1, *pwm_duty1;
```

```
FILE *pwm_polarity1, *pwm_run1;

//define pin variables for channel 1
int mux_mode1 = 0x006, request1 = 1, freq1 = 100, duty1 = 50;
int polarity1 = 1, run1 = 1;

ofp_gpmc_a3 = fopen("/sys/kernel/debug/omap_mux/gpmc_a3", "w");
if(ofp_gpmc_a3 == NULL) {printf("Unable to open gpmc_a3.\n");}
fseek(ofp_gpmc_a3, 0, SEEK_SET);
fprintf(ofp_gpmc_a3, "0x%02x", mux_mode1);
fflush(ofp_gpmc_a3);

pwm_req1 = fopen("/sys/class/pwm/ehrpwm.1:1/request", "w");
if(pwm_req1 == NULL) {printf("Unable to open pwm 1 request.\n");}
fseek(pwm_req1, 0, SEEK_SET);
fprintf(pwm_req1, "%d", request1);
fflush(pwm_req1);

pwm_freq1 = fopen("/sys/class/pwm/ehrpwm.1:1/period_freq", "w");
if(pwm_freq1 == NULL) {printf("Unable to open pwm 1 frequency.\n");}
fseek(pwm_freq1, 0, SEEK_SET);
fprintf(pwm_freq1, "%d", freq1);
fflush(pwm_freq1);

pwm_duty1 = fopen("/sys/class/pwm/ehrpwm.1:1/duty_percent", "w");
if(pwm_duty1 == NULL) {printf("Unable to open pwm 1 duty cycle.\n");}
fseek(pwm_duty1, 0, SEEK_SET);
fprintf(pwm_duty1, "%d", duty1);
fflush(pwm_duty1);

pwm_polarity1 = fopen("/sys/class/pwm/ehrpwm.1:1/polarity", "w");
if(pwm_polarity1 == NULL) {printf("Unable to open pwm 1 polarity.\n");}
fseek(pwm_polarity1, 0, SEEK_SET);
fprintf(pwm_polarity1, "%d", polarity1);
fflush(pwm_polarity1);

pwm_run1 = fopen("/sys/class/pwm/ehrpwm.1:1/run", "w");
if(pwm_run1 == NULL) {printf("Unable to open pwm 1 run.\n");}
```

```
while(1)
  {
  //read analog sensors
  fseek(ifp_ain1, 0, SEEK_SET);
  fscanf(ifp_ain1, "%f", &ain0_value);
  printf("%f\n", ain0_value);

  fseek(ifp_ain2, 0, SEEK_SET);
  fscanf(ifp_ain2, "%f", &ain1_value);
  printf("%f\n", ain1_value);

  fseek(ifp_ain3, 0, SEEK_SET);
  fscanf(ifp_ain3, "%f", &ain2_value);
  printf("%f\n", ain2_value);

  //implement truth to determine robot turns

  //no walls present - continue straight ahead
  if((ain0_value < threshold)&&(ain1_value < threshold)&&
     (ain2_value < threshold))
    {
    run0 = 1;  run1 = 1;          //both motors on
    fseek(pwm_run0, 0, SEEK_SET);
    fprintf(pwm_run0, "%d", run0);
    fflush(pwm_run0);

    fseek(pwm_run1, 0, SEEK_SET);
    fprintf(pwm_run1, "%d", run1);
    fflush(pwm_run1);
    }
/*
  else if(...)
    {

    :
    insert other cases
    :
    }
```

```
*/
  }

fclose(ifp_ain1);
fclose(ifp_ain2);
fclose(ifp_ain3);

fclose(ofp_gpmc_a2);
fclose(pwm_req0);
fclose(pwm_freq0);
fclose(pwm_duty0);
fclose(pwm_polarity0);
fclose(pwm_run0);

fclose(ofp_gpmc_a3);
fclose(pwm_req1);
fclose(pwm_freq1);
fclose(pwm_duty1);
fclose(pwm_polarity1);
fclose(pwm_run1);

return 1;
}
//*********************************************************************
```

BeagleBone Black–Linux 3.8

```
//*********************************************************************
//dagu.cpp
//
//Before the right interfaces show up, you must have performed:
//   echo cape-bone-iio > /sys/devices/bone_capemgr.*/slots
//   echo am335x_pwm > /sys/devices/bone_capemgr.*/slots
//   echo bone_pwm_P9_14 > /sys/devices/bone_capemgr.*/slots
//   echo bone_pwm_P9_16 > /sys/devices/bone_capemgr.*/slots
//
//Note:
// * The .11 in helper.11 is likely to change
// * The .12 and .15 in the PWM entries are likely to change
// * The .2 in ocp might change
//*********************************************************************
```

```c
#include <stdio.h>
#include <stddef.h>
#include <time.h>
#include <math.h>

#define  output "out"
#define  input  "in"

int  main (void)
{
//wall detection threshold
double threshold = 1100.0;                    //experimentally determined

//configure adc channels
//define file handles for adc related files
FILE *ifp_ain0, *ifp_ain1, *ifp_ain2;
float ain0_value, ain1_value, ain2_value;

//open adc related files for access to ain0, 1 and 2
ifp_ain0 = fopen("/sys/module/bone_iio_helper/drivers/
platform:bone-iio-helper/helper.11/AIN0", "r");
if (ifp_ain0 == NULL) {printf("Unable to ain0.\n");}

ifp_ain1 = fopen("/sys/module/bone_iio_helper/drivers/
platform:bone-iio-helper/helper.11/AIN1", "r");
if (ifp_ain1 == NULL) {printf("Unable to ain1.\n");}

ifp_ain2 = fopen("/sys/module/bone_iio_helper/drivers/
platform:bone-iio-helper/helper.11/AIN2", "r");
if (ifp_ain2 == NULL) {printf("Unable to ain2.\n");}

//configure pwm channels 0 and 1
//define file handles for channel 0 - EHRPWM1A (P9, pin 14)
//designated as ehrpwm.1:0
FILE *pwm_period0, *pwm_duty0;
FILE *pwm_polarity0, *pwm_run0;

//define pin variables for channel 0
int period0 = 500000, duty0 = 250000;
```

```
int polarity0 = 1, run0 = 1;

pwm_period0 = fopen("/sys/devices/ocp.2/pwm_test_P9_14.12/period", "w");
if(pwm_period0 == NULL) {printf("Unable to open pwm 0 period.\n");}
fseek(pwm_period0, 0, SEEK_SET);
fprintf(pwm_period0, "%d", period0);
fflush(pwm_period0);

pwm_duty0 = fopen("/sys/devices/ocp.2/pwm_test_P9_14.12/duty", "w");
if(pwm_duty0 == NULL) {printf("Unable to open pwm 0 duty cycle.\n");}
fseek(pwm_duty0, 0, SEEK_SET);
fprintf(pwm_duty0, "%d", duty0);
fflush(pwm_duty0);

pwm_polarity0 = fopen("/sys/devices/ocp.2/pwm_test_P9_14.12/polarity", "w");
if(pwm_polarity0 == NULL) {printf("Unable to open pwm 0 polarity.\n");}
fseek(pwm_polarity0, 0, SEEK_SET);
fprintf(pwm_polarity0, "%d", polarity0);
fflush(pwm_polarity0);

pwm_run0 = fopen("/sys/devices/ocp.2/pwm_test_P9_14.12/run", "w");
if(pwm_run0 == NULL) {printf("Unable to open pwm 0 run.\n");}

//define file handles for channel 1 - EHRPWM1B (P9, pin 16)
//designated as ehrpwm.1:1
FILE *pwm_period1, *pwm_duty1;
FILE *pwm_polarity1, *pwm_run1;

//define pin variables for channel 1
int period1 = 500000, duty1 = 250000;
int polarity1 = 1, run1 = 1;

pwm_period1 = fopen("/sys/devices/ocp.2/pwm_test_P9_16.15/period", "w");
if(pwm_period1 == NULL) {printf("Unable to open pwm 1 period.\n");}
fseek(pwm_period1, 0, SEEK_SET);
fprintf(pwm_period1, "%d", period1);
fflush(pwm_period1);

pwm_duty1 = fopen("/sys/devices/ocp.2/pwm_test_P9_16.15/duty", "w");
```

```
if(pwm_duty1 == NULL) {printf("Unable to open pwm 1 duty cycle.\n");}
fseek(pwm_duty1, 0, SEEK_SET);
fprintf(pwm_duty1, "%d", duty1);
fflush(pwm_duty1);


pwm_polarity1 = fopen("/sys/devices/ocp.2/pwm_test_P9_16.15/polarity", "w");
if(pwm_polarity1 == NULL) {printf("Unable to open pwm 1 polarity.\n");}
fseek(pwm_polarity1, 0, SEEK_SET);
fprintf(pwm_polarity1, "%d", polarity1);
fflush(pwm_polarity1);


pwm_run1 = fopen("/sys/devices/ocp.2/pwm_test_P9_16.15/run", "w");
if(pwm_run1 == NULL) {printf("Unable to open pwm 1 run.\n");}


while(1)
  {
  //read analog sensors
  fseek(ifp_ain0, 0, SEEK_SET);
  fscanf(ifp_ain0, "%f", &ain0_value);
  printf("%f\n", ain0_value);

  fseek(ifp_ain1, 0, SEEK_SET);
  fscanf(ifp_ain1, "%f", &ain1_value);
  printf("%f\n", ain1_value);

  fseek(ifp_ain2, 0, SEEK_SET);
  fscanf(ifp_ain2, "%f", &ain2_value);
  printf("%f\n", ain2_value);

  //implement truth to determine robot turns

  //no walls present - continue straight ahead
  if((ain0_value < threshold)&&(ain1_value < threshold)&&
     (ain2_value < threshold))
    {
    run0 = 1;  run1 = 1;          //both motors on
    fseek(pwm_run0, 0, SEEK_SET);
    fprintf(pwm_run0, "%d", run0);
```

```
      fflush(pwm_run0);

      fseek(pwm_run1, 0, SEEK_SET);
      fprintf(pwm_run1, "%d", run1);
      fflush(pwm_run1);
      }
/*
  else if(...)
    {

    :
    insert other cases
    :
    }
*/
  }

fclose(ifp_ain0);
fclose(ifp_ain1);
fclose(ifp_ain2);

fclose(pwm_period0);
fclose(pwm_duty0);
fclose(pwm_polarity0);
fclose(pwm_run0);

fclose(pwm_period1);
fclose(pwm_duty1);
fclose(pwm_polarity1);
fclose(pwm_run1);

return 1;
}
//****************************************************************
```

6.6   APPLICATION 4: PORTABLE IMAGE PROCESSING ENGINE

Image processing is a fascinating field. It is the process of extracting useful information from an image. Image processing operations typically requires the application of a series of simple operations to an image. Each operator is sequentially swept over an image to accomplish a specific task. BeagleBone, with its 1 GHz clock speed, is ideally suited for image processing applications. In fact, BeagleBone equipped with a small keyboard and the Circuitco LCD7 liquid crystal display may be viewed as a portable image processing engine. In this section, we provide a brief introduction to image processing, the layout for a BeagleBone portable image processing engine, a brief introduction to the OpenCV image processing library and conclude with an example. The "stache cam" example uses a variety of OpenCV features to perform face recognition and place a moustache on the face. This appears to be a fun application (it is); however, it also forms the basis for a variety of assistive technology applications such as pupil tracking.

6.6.1   BRIEF INTRODUCTION TO IMAGE PROCESSING

A basic image processing system is shown in Figure 6.19. It consists of a camera equipped with a lens system, a frame grabber and a host processor to perform the image processing task. The camera captures images of objects onto a two–dimensional light sensitive array. Each element in the array is termed a picture element or pixel as shown in Figure 6.20. The camera "snaps" an image of a scene as determined by the frame rate of the camera. Typical frames rates include 30 and 60 frames per second (fps). Other high speed frame rates are available. Also, slower frame rates are available in high resolution cameras. Spatial resolution refers to the number of pixels within the imaging array. Although higher resolution will resolve finer object detail, it comes at the expense of increased computational cost.

Cameras are available in black and white or color. Usually a black and white camera will register the shade or gray scale of each pixel as a single byte. A gray scale value of 0 is assigned to black while 255 is used to designate white. Shades between black and white are assigned gray scale values between 0 and 255.

Various color schemes may be used to represent an image including red–green–blue (R–G–B), hue–saturation–intensity (H–S–I), etc. They require three bytes to represent a pixel in a specific color scheme.

Image processing operations typically requires the application of a series of simple operations to an image. Each operator is sequentially swept over an image to accomplish a specific task a shown in Figure 6.21. For example, the low pass filter kernel may be exhaustively applied to each pixel in an image to smooth image features. To apply a filter operator, the filter coefficients are multiplied by the image coefficients and then summed. The resulting value becomes the new pixel value at that image pixel location. The filter operator is exhaustively applied to every pixel in the image. Although the operations are quite straightforward, they require a substantial amount of processor power due to the sheer number of calculations that must be accomplished in a timely manner [Galbiati, Gonzalez

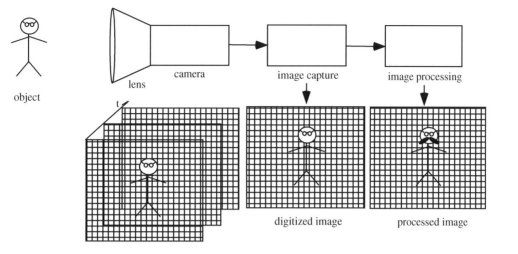

Figure 6.19: Image processing system.

and Woods]. As previously mentioned BeagleBone is well suited for a portable image processing engine.

6.6.2 OPENCV COMPUTER VISION LIBRARY

The OpenCV Library is an open source computer vision library. The library is written in C and C++ and runs on a variety of operating systems including Linux, Windows and Mac OS X. The library allows a designer to rapidly prototype an image processing application. Bradski and Kaehler provides an excellent tutorial on this library [Bradski and Kaehler]. With its Linux–based operating system, BeagleBone is ideally suited to host the OpenCV library. We illustrate the power and application of the OpenCV library with the "Stache Cam" application.

6.6.3 STACHE CAM

In this section we illustrate the power and application of the OpenCV library with the "Stache Cam" application. The requirements for this system are straightforward. A video camera is used to capture images of various faces. Functions within the Open CV library are used to capture the image and automatically place a moustache on the face as shown in Figure 6.22.

The system hardware consists of a camera (Playstation PS3 Eye), BeagleBone, the LCD3 BeagleBone Cape, and the BeagleBone Battery cape for portability. Linux provides a USB driver for the PS3 Eye camera. Different system hardware components are shown in Figure 6.23.

Stache cam UML activity diagram
The UML activity diagram is provided in Figure 6.24.

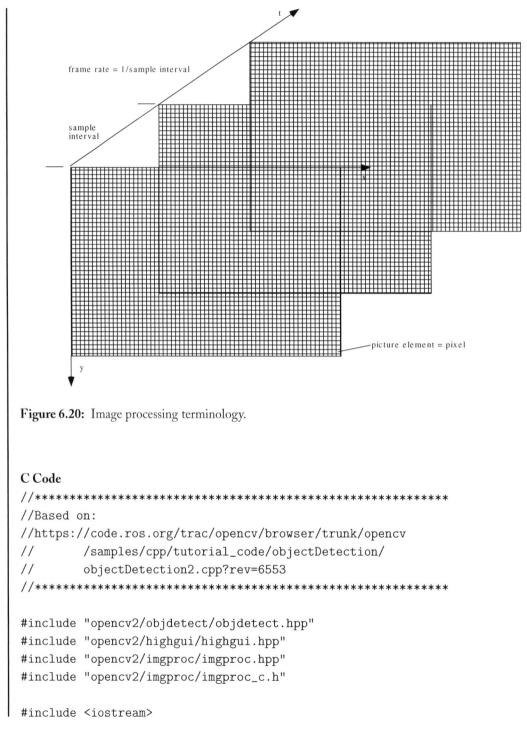

Figure 6.20: Image processing terminology.

C Code
```
//***********************************************************
//Based on:
//https://code.ros.org/trac/opencv/browser/trunk/opencv
//        /samples/cpp/tutorial_code/objectDetection/
//        objectDetection2.cpp?rev=6553
//***********************************************************

#include "opencv2/objdetect/objdetect.hpp"
#include "opencv2/highgui/highgui.hpp"
#include "opencv2/imgproc/imgproc.hpp"
#include "opencv2/imgproc/imgproc_c.h"

#include <iostream>
```

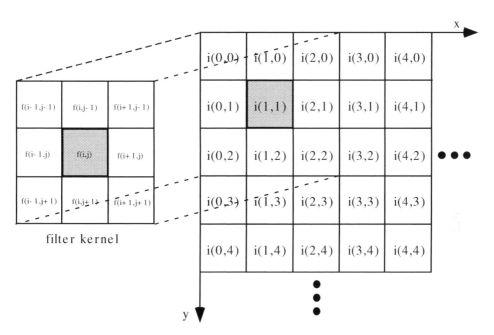

a) filter operation

b) filter operators

Figure 6.21: Image processing filters.

Figure 6.22: BeagleBone "stache" cam.

Figure 6.23: BeagleBone "stache" cam hardware.

```c
#include <stdio.h>

using namespace std;
using namespace cv;

string copyright = "\

IMPORTANT:READ BEFORE DOWNLOADING, COPYING, INSTALLING
OR USING.\n\\n\
By downloading, copying, installing or using the software
you agree to this license.\n\
If you do not agree to this license, do not download,
```

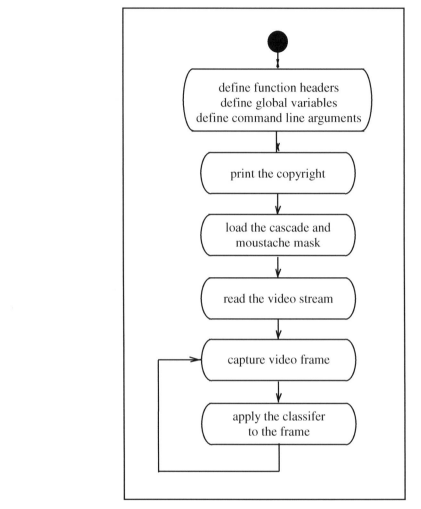

Figure 6.24: BeagleBone "stache" cam UML activity diagram.

```
install, copy or use the software.\n\
\n\
\n\
                        License Agreement\n\
            For Open Source Computer Vision Library\n\
\n\
Copyright(C)2000-2008, Intel Corporation, all rights reserved.\n\
Copyright(C)2008-2011, Willow Garage Inc., all rights reserved.\n\
```

```
Copyright(C)2012, Texas Instruments, all rights reserved.\n\
Third party copyrights are property of their respective owners.\n\\n\
Redistribution and use in source and binary forms, with or without\n\
modification, are permitted provided that the following conditions\n\
are met:\n\
\n\
   *Redistributions of source code must retain the above \n\
    copyright notice, this list of conditions and the following\n\
    disclaimer.\n\
\n\
   *Redistributions in binary form must reproduce the above \n\
    copyright notice, this list of conditions and the following \n\
    disclaimer in the documentation and/or other materials provided \n\
    with the distribution.\n\
\n\
   *The name of the copyright holders may not be used to endorse or \n\
    promote products derived from this software without specific \n\
    prior written permission.\n\
\n\
This software is provided by the copyright holders and contributors\n\
\"as is\" and any express or implied warranties, including, but not\n\
limited to, the implied warranties of merchantability and fitness\n\
for a particular purpose are disclaimed.\n\
In no event shall the Intel Corporation or contributors be liable\n\
for any direct, indirect, incidental, special, exemplary, or\n\
consequential damages (including, but not limited to, procurement\n\
of substitute goods or services; loss of use, data, or profits; or\n\
business interruption) however caused and on any theory of liability,\n\
whether in contract, strict liability, or tort (including negligence\n\
or otherwise) arising in any way out of the use of this software, even\n\
if advised of the possibility of such damage.\n\
\n";

/** Function Headers */
void detectAndDisplay(Mat frame);

/** Global variables */
String face_cascade_name = "lbpcascade_frontalface.xml";
CascadeClassifier face_cascade;
```

```
string window_name = "stache - BeagleBone OpenCV demo";
IplImage* mask = 0;


/** Command-line arguments */
int numCamera = -1;
const char* stacheMaskFile = "stache-mask.png";
int scaleHeight = 6;
int offsetHeight = 4;
int camWidth = 0;
int camHeight = 0;
int camFPS = 0;


/*** @function main */
int main(int argc, const char** argv)
  {
  CvCapture* capture;
  Mat frame;

  if(argc > 1) numCamera = atoi(argv[1]);
  if(argc > 2) stacheMaskFile = argv[2];
  if(argc > 3) scaleHeight = atoi(argv[3]);
  if(argc > 4) offsetHeight = atoi(argv[4]);
  if(argc > 5) camWidth = atoi(argv[5]);
  if(argc > 6) camHeight = atoi(argv[6]);
  if(argc > 7) camFPS = atoi(argv[7]);

  //-- 0. Print the copyright
  cout << copyright;

  //-- 1. Load the cascade
  if( !face_cascade.load(face_cascade_name) ){ printf("--(!)Error
      loading\n"); return -1; };

  //-- 1a.
Load the mustache mask
  mask = cvLoadImage(stacheMaskFile);
  if(!mask) { printf("Could not load %s\n", stacheMaskFile); exit(-1); }

  //-- 2. Read the video stream
```

```
capture = cvCaptureFromCAM(numCamera);
if(camWidth) cvSetCaptureProperty(capture, CV_CAP_PROP_FRAME_WIDTH,
camWidth);
if(camHeight) cvSetCaptureProperty(capture, CV_CAP_PROP_FRAME_HEIGHT,
camHeight);
if(camFPS) cvSetCaptureProperty(capture, CV_CAP_PROP_FPS, camFPS);
if(capture)
  {
  while(true)
    {
    frame = cvQueryFrame(capture);

    //-- 3. Apply the classifier to the frame
    try
      {
      if(!frame.empty())
        {
        detectAndDisplay( frame );
        }
      else
        {
        printf(" --(!) No captured frame---Break!\n"); break;
        }
      int c = waitKey(10);
      if( (char)c == 'c' )
        {
        break;
        }
      }
    catch(cv::Exception e)
      {
      }
    }
  }
  return 0;
}

//****************************************************
/*** @function detectAndDisplay */
```

```
//****************************************************

void detectAndDisplay(Mat frame)
  {
  std::vector<Rect> faces;
  Mat frame_gray;

  cvtColor(frame, frame_gray, CV_BGR2GRAY);
  equalizeHist(frame_gray, frame_gray);

  //-- Detect faces
  face_cascade.detectMultiScale(frame_gray,
  faces, 1.1, 2, 0, Size(80, 80));

  for(int i = 0; i < faces.size(); i++)
    {
    //-- Scale and apply mustache mask for each face
    Mat faceROI = frame_gray(faces[i]);
    IplImage iplFrame = frame;
    IplImage *iplMask = cvCreateImage(cvSize(faces[i].width,
                              faces[i].height/scaleHeight),
                              mask->depth, mask->nChannels);
    cvSetImageROI(&iplFrame, cvRect(faces[i].x,
      faces[i].y + (faces[i].height/scaleHeight)*offsetHeight,
      faces[i].width, faces[i].height/scaleHeight));


    cvResize(mask, iplMask, CV_INTER_LINEAR);
    cvSub(&iplFrame, iplMask, &iplFrame);
    cvResetImageROI(&iplFrame);
    }

//-- Show what you got
flip(frame, frame, 1);
imshow(window_name, frame);
}
//****************************************************
```

6.7   SUMMARY

In the early chapters of this book, we examined the Bonescript environment as a user–friendly tool to rapidly employ BeagleBone features right out of the box. In this chapter we revisited Bonescript and demonstrated its power as a rapid prototyping tool to develop complex, processor–based systems employing multiple BeagleBone subsystems. We carefully chose each of the projects to illustrate how BeagleBone may be used in a variety of project areas including instrumentation intensive applications (weather station), in assistive and educational technology applications (Speak–and–Spell), in motor and industrial control applications (Dagu robot), and calculation intensive image processing applications (moustache cam).

6.8   REFERENCES

- Kelley, A. and I. Pohl. *A Book on C — Programming in C.* 4th edition. Boston, MA: Addison Wesley, 1998. Print.

- Vander Veer, Emily. *JavaScript for Dummies.* Hoboken, NJ: Wiley Publishing, Inc., 4th edition, 2005. Print.

- Pollock, John. *JavaScript.* New York, NY: McGraw Hill, 3rd edition, 2010. Print.

- Kiessling, Manuel. *The Node Beginner Guide: A Comprehensive Node.js Tutorial.* 2012. Print.

- Hughes–Croucher, Tom and Mike Wilson. *Node Up and Running.* Sebastopol, CA: O'Reilly Media, Inc., 2012. Print.

- von Hagen, W. *Ubuntu Linux Bible.* Indianapolis, IN: Wiley Publishing, Inc., 2007. Print.

- Coley, Gerald. *BeagleBone Rev A6 Systems Reference Manual.* Revision 0.0, May 9, 2012, beagleboard.org `www.beaglebord.org`

- Barrett, Steven and Daniel Pack. *Embedded Systems Design and Applications with the 68HC12 and HCS12.* Upper Saddle River, NJ: Pearson Prentice Hall, 2005. Print.

- Barrett, Steven and Daniel Pack. *Processors Fundamentals for Engineers and Scientists.* Morgan and Claypool Publishers, 2006. `www.morganclaypool.com`

- Barrett, Steven and Daniel Pack. *Atmel AVR Processor Primer Programming and Interfacing.* Morgan and Claypool Publishers, 2008. `www.morganclaypool.com`

- Galbiati, L. *Machine Vision and Digital Image Processing Fundamentals.* Englewood Cliffs, NJ: Prentice Hall, 1990. Print.

- Gonzalez, R.C. and R.E. Woods. *Digital Image Processing.* 3rd edition, Upper Saddle River, NJ: Prentice Hall, 2008.

- Bradski, D. and A. Kaehler. *Learning OpenCV: Computer Vision with the OpenCV Library.* Sebastopol, CA: O'Reilly, 2008.

6.9 CHAPTER EXERCISES

1. Construct the UML activity diagrams for all functions related to the weather station.

2. It is desired to updated weather parameters every 15 minutes. Write a function to provide a 15 minute delay in Bonescript and C.

3. Add one of the following sensors to the weather station:

   - anemometer
   - barometer
   - hygrometer
   - rain gauge
   - thermocouple

   You will need to investigate background information on the selected sensor, develop an interface circuit for the sensor, and modify the weather station code.

4. Complete the control algorithm for the weather station to convert the reading from the LM34 temperature sensor and display its value and wind direction.

5. The Blinky 602A robot under microcontroller control abruptly starts and stops when PWM is applied. Modify the algorithm to provide the capability to gradually ramp up (and down) the motor speed.

6. Modify the Blinky 602A circuit and microcontroller code such that the maximum speed of the robot is set with an external potentiometer.

7. Modify the Blinky 602A circuit and microcontroller code such that the IR sensors are only asserted just before a range reading is taken.

8. Add the following features to the Blinky 602A platform:

   - Line following capability (Hint: Adapt the line following circuitry onboard the Blinky 602A to operate with the BeagleBone.)
   - Two way robot communications (use the IR sensors already aboard)
   - Voice output (Hint: Use the SP0–512 speech synthesis chip.)

CHAPTER 7

Where to from here?

Objectives: After reading this chapter, the reader should be able to do the following:

- View this book as simply the beginning of the journey in using BeagleBone in a wide variety of applications.

- Describe the wide variety of software libraries and other resources available to the BeagleBone user.

- Appreciate the advantages in becoming an active member of the BeagleBoard.org community.

7.1 OVERVIEW

Reaching the last chapter of the book, you might be breathing a sigh of relief thinking I've completed the book. I'm at the end. Quite the contrary, this book is merely the beginning of your journey of using BeagleBone in a wide variety of applications.

In this chapter we provide a brief review of a number of software libraries and other resources available to a BeagleBone user. These resources allow access to a wide array of features to extend the capabilities of BeagleBone. We conclude with an invitation to become an active member of the BeagleBoard.org community.

7.2 SOFTWARE LIBRARIES

7.2.1 OPENCV

The OpenCV Library is an open source computer vision library. The library is written in C and C++ and runs on a variety of operating systems including Linux, Windows and Mac OS X. The library allows a designer to rapidly prototype an image processing application. Bradski and Kaehler provides an excellent tutorial on this library [Bradski and Kaehler]. With a Linux–based operating system, BeagleBone is ideally suited to host the OpenCV library.

7.2.2 QT

Qt is a C++ library that provides for the rapid development of user–friendly graphical user interfaces of GUIs. Qt readily executes on Windows, Unix, MacOS X and Linux–based embedded systems. The Qt library allows GUIs that employ buttons, scroll bars, etc. The library employs the concept of signals and slots to link an event to a desired response [Dalheimer]. Several excellent sources on the Qt library are listed at the end of the chapter.

7.2.3 KINECT

Kinect is the motion sensing input device developed for the Xbox video game console. The Kinect allows natural user movements such as gestures or spoken commands to interact with a game. A Kinect library is available for BeagleBone operating in Linux. A good introduction to Kinect is provided by the references listed at the end of the chapter.

7.3 ADDITIONAL RESOURCES

In this section we provide pointers to a series of user–groups and other resources available to a BeagleBone user.

7.3.1 OPENROV

Earlier in the book, we introduced an underwater remote operated vehicle (ROV) project. There are a number of groups dedicated to ROV development as a way of reaching the next generation of engineers and scientists. A brief description of the groups is provided below.

- OpenROV is a do–it–yourself (DIY) group dedicated to underwater robots for exploration and adventure. The group includes amateur and professional ROV builders and operators from over 50 countries who have a passion for exploring the deep [www.openrov.com]. The OpenROV community has a BeagleBone Cape available for controlling an ROV as shown in Figure 7.1.

- SeaPerch is an underwater robotics program to equip educators and students with resources to build an underwater Remotely Operated Vehicle (ROV). There are a number of excellent texts on underwater ROV development listed at the end of the chapter [www.seaperch.org].

7.3.2 NINJA BLOCKS

Ninja Blocks is an innovative method to sense the environment and control hardware within the home. Ninja Blocks is based on an open hardware concept where hardware, software and application information is openly shared among the Ninja Blocks community. The basic Ninja Block unit is illustrated in Figure 7.2. The basic unit includes the following components [www.ninjablocks.com]:

- wireless motion sensor

- wireless door/window contact sensor

- wireless button

- wireless temperature and humidity sensor

- Ninja Block equipped with a BeagleBone and an Arduino processor

Figure 7.1: The OpenROV BeagleBone Cape [www.openrov.com].

- USB Wi–Fi module

- Ethernet Cable

- 5 VDC, 3 Amp power supply with connectors

    The basic unit allows ease of interface to a wide variety of devices.

7.3.3   BEAGLEBOARD.ORG RESOURCES

The BeagleBoard.org community has many members. What we all have in common is the desire to put processing power in the hands of the next generation of users. BeagleBoard.org, with
Texas Instruments' support, embraced the open source concept with the development and release
of BeagleBone in late 2011. Their support will insure the BeagleBone project will be sustainable.
BeagleBoard.org partnered with circuitco (www.circuitco.com) to produce BeagleBone and its
associated Capes. The majority of the Capes have been designed and fabricated by circuitco. Clint
Cooley, President of circuitco, is most interested in helping users develop and produce their own

Figure 7.2: Ninja Blocks [`www.ninjablocks.com`].

ideas for BeagleBone Capes. Texas Instruments has also supported the BeagleBoard.org community by giving Jason Kridner the latitude to serve as the open platform technologist and evangelist for the BeagleBoard.org community. The most important members of the community are the BeagleBoard and Bone users. Our ultimate goal is for the entire community to openly share their successes and to encourage the next generation of STEM practitioners.

7.3.4   CONTRIBUTING TO BONESCRIPT

It is important to emphasize that Bonescript is an open source programming environment. We are counting on the user community to expand the features of Bonescript. If there is a feature you need, please develop it and share it with the BeagleBoard.org community. This is easily done by submitting a pull request to `www.github.com/jadonk/bonescript`.

7.4   SUMMARY

In this chapter we provided a brief review of a number of software libraries and other resources available to the BeagleBone user. These resources allow access to a wide array of features to extend the capabilities of BeagleBone. We concluded with an invitation to become an active member of the BeagleBoard.org community.

7.5 REFERENCES

- Bradski, D. and A. Kaehler. *Learning OpenCV: Computer Vision with the OpenCV Library.* Sebastopol, CA: O'Reilly, 2008.

- Thelin, J. *Foundations of Qt Development.* Berkerly, CA: Apress, 2007.

- Dalheimer, M. *Programming with Qt.* Sebastopol, CA: O'Reilly, 2002.

- Miles, R. *Start Here! Learn the Kinect API.* Sebastopol, CA: O'Reilly, 2012.

- Kean, S., J. Hall and P. Perry. *Meet the Kinect: An Introduction to Programming Natural User Interfaces.* Berkeley, CA: Apress, 2012.

- *Seaperch,* `www.seaperch.com`

- Bohm, H. and V. Jensen. *Build Your Own Underwater Robot and Other Wet Projects.* Monterey, CA: Marine Advanced Technology Center, 2012.

- Moore, S., H. Bohm, and V. Jensen. *Underwater Robotics: Science, Design and Fabrication.*

- *OpenROV: Open–source Underwater Robots for Exploration and Education* , `www.openrov.com`

7.6 CHAPTER EXERCISES

1. Construct a personal plan on how you will improve your BeagleBone and Linux operating system skills.

2. Develop three new features for the Bonescript environment and submit them to the Beagle-Board.org community.

APPENDIX A

Bonescript functions

Bonescript Environment

Analog input/output

function name	Description
var1 = analogRead(pin_name);	**Description:** Performs analog-to-digital conversion on voltage at specified pin. Analog voltage may range from 0 to 1.8 VDC. **arguments:** pin_name **returns:** Normalized value from 0 .. 1 corresponding to 0 .. 1.8 VDC.
analogWrite(pin_name, analog_value);	**Description:** Delivers analog level to specified pin via 1 kHz pulse width modulated signal. Analog level specified as normalized value from 0..1 corresponding to 0 to 1.8 VDC. **arguments:** pin_name, logic_level (0 .. 1). **returns:** None.

Figure A.1: Bonescript analog input and output functions.

Bonescript Environment

Digital input/output

function name	Description
pinMode(pin_name, direction);	**Description:** Sets digital pin direction (INPUT or OUTPUT). **arguments:** pin_name, direction (INPUT or OUTPUT) **returns:** None.
getPinMode(pin_name);	**Description:** Reports status, parameters of selected pin **arguments:** pin_name **returns:** parameters of selected pin
digitalWrite(pin_name, logic_level);	**Description:** Sets digital output pin logic level (HIGH or LOW). **arguments:** pin_name, logic_level (HIGH or LOW) **returns:** None.
var1 = digitalRead(pin_name);	**Description:** Reads digital input pin logic level (HIGH or LOW). **arguments:** pin_name **returns:** Logic level of specified pin (HIGH or LOW)
shiftOut(spi_dataPin, spi_clockPin, bitOrder, value);	**Description:** Provides Serial Peripheral Interface (SPI) data transmission **arguments:** pin for SPI data, pin for SPI clock, bit order (LSBFIRST, MSBFIRST), data for transmission **returns:** None.

Figure A.2: Bonescript digital input and output functions.

Bonescript Environment

Bit and Byte Operators

function name	Description
return_byte = lowByte(value);	**Description:** Returns the lower (least significant) byte of the value. The value may be of any type. **arguments:** The value may be of any type. **returns:** The low order, least significant byte.
bit_value = bitRead(number, bit_position);	**Description:** Returns the logic value (1 or 0) of the specified bit position in the number. **arguments:** The number to be evaluated and the desired bit position. **returns:** The logic value (1 or 0).

bitWrite(number,
      bit_position,
      bit_value);
	Description: Writes the specified bit position with a bit value (0 or 1) for the specified number.
	
	arguments: The number to be written to, the desired bit position and the desired value (0 or 1).
	
	returns: None.

bitSet(number, bit_position);	**Description:** Sets the specified bit position to logic high (1) for the specified number. **arguments:** The number to be written to and the desired bit position. **returns:** None.
bitClear(number, bit_position);	**Description:** Clears the specified bit position to logic low (0) for the specified number. **arguments:** The number to be written to and the desired bit position. **returns:** None.
bit_value = bit(n);	**Description:** Returns 2^n as the bit value. **arguments:** The bit position (n). **returns:** The bit value.

Figure A.3: Bonescript bit and byte operators.

Bonescript Environment

Interrupts

function name	Description
attachInterrupt(input pin, ISR_name, trigger);	**Description:** Specifies input pin, interrupt trigger source, and interrupt service routine (ISR) name **arguments:** interrupt pin, trigger source RISING, FALLING, CHANGE), and interrupt service routine name **returns:** None.
detachInterrupt(pin);	**Description:** Detaches interrupt service from specified pin. **arguments:** Interrupt pin **returns:** None.

Constants

OUTPUT INPUT LOW HIGH RISING FALLING CHANGE

Figure A.4: Bonescript interrupt operators and constants.

APPENDIX B

LCD interface for BeagleBone in C

B.1 BEAGLEBONE ORIGINAL – LINUX 3.2

Provided in Figure B.1 are the structure chart, UML activity diagrams and connection diagram for a Sparkfun LCD–09052 basic 16 x 2 character liquid crystal display. Note the LCD operates at 3.3 VDC which is compatible with BeagleBone.

Provided below is the C code for the BeagleBone original operating under Linux–3.2.

```cpp
//****************************************************************
//lcd3.cpp
//****************************************************************

#include <stdio.h>
#include <stdlib.h>
#include <stddef.h>
#include <time.h>
#include <string.h>
#include <sys/time.h>

#define   output  "out"
#define   input   "in"

//function prototypes
void configure_LCD_interface(void);
void LCD_putcommand(unsigned char);
void LCD_putchar(unsigned char);
void LCD_print(unsigned int, char *string);
void delay_us(int);
void LCD_init(void);

//LCD_RS: bone.P8_12 LCD Register Set (RS) control: GPIO1_12: pin-44
FILE *ofp_gpmc_ad12, *export_44, *gpio44_value, *gpio44_direction;
```

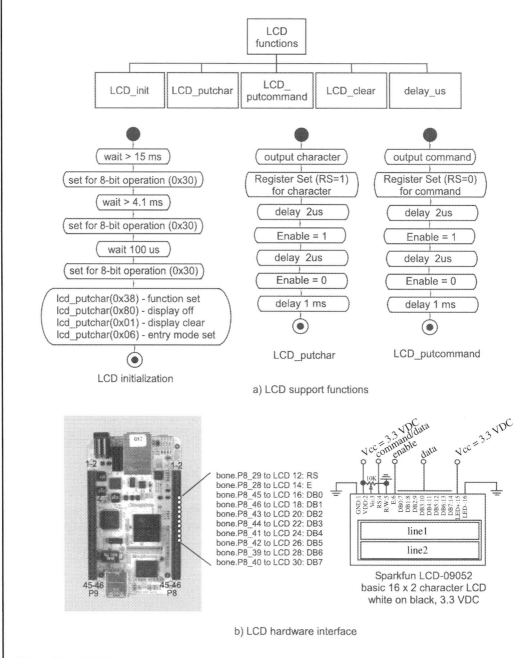

b) LCD hardware interface

Figure B.1: LCD support data.

```
int mux_mode_44 = 0x8007, pin_number_44 = 44, logic_status_44 = 1;
char* pin_direction_44 = output;

//LCD_E:   bone.P8_14: LCD Enable En (E): GPIO0_26: pin-26
FILE *ofp_gpmc_ad10, *export_26, *gpio26_value, *gpio26_direction;
int mux_mode_26 = 0x8007, pin_number_26 = 26, logic_status_26 = 1;
char* pin_direction_26 = output;

//LCD_DB0: bone.P8_16: LCD Data line DB0: GPIO1_14: pin-46
FILE *ofp_gpmc_ad14, *export_46, *gpio46_value, *gpio46_direction;
int mux_mode_46 = 0x8007, pin_number_46 = 46, logic_status_46 = 1;
char* pin_direction_46 = output;

//LCD_DB1: bone.P8_18: LCD Data line DB1: GPIO2_1: pin-65
FILE *ofp_gpmc_clk, *export_65, *gpio65_value, *gpio65_direction;
int mux_mode_65 = 0x8007, pin_number_65 = 65, logic_status_65 = 1;
char* pin_direction_65 = output;

//LCD_DB2: bone.P8_20: LCD Data line DB2: GPIO1_31: pin-63
FILE *ofp_gpmc_csn2, *export_63, *gpio63_value, *gpio63_direction;
int mux_mode_63 = 0x8007, pin_number_63 = 63, logic_status_63 = 1;
char* pin_direction_63 = output;

//LCD_DB3: bone.P8_22: LCD Data line DB3: GPIO1_5: pin-37
FILE *ofp_gpmc_ad5, *export_37, *gpio37_value, *gpio37_direction;
int mux_mode_37 = 0x8007, pin_number_37 = 37, logic_status_37 = 1;
char* pin_direction_37 = output;

//LCD_DB4: bone.P8_24: LCD Data line DB4: GPIO1_1: pin-33
FILE *ofp_gpmc_ad1, *export_33, *gpio33_value, *gpio33_direction;
int mux_mode_33 = 0x8007, pin_number_33 = 33, logic_status_33 = 1;
char* pin_direction_33 = output;

//LCD_DB5: bone.P8_26: LCD Data line DB5: GPIO1_29: pin-61
FILE *ofp_gpmc_csn0, *export_61, *gpio61_value, *gpio61_direction;
int mux_mode_61 = 0x8007, pin_number_61 = 61, logic_status_61 = 1;
char* pin_direction_61 = output;

//LCD_DB6: bone.P8_28: LCD Data line DB6: GPIO2_24: pin-88
```

```
FILE *ofp_lcd_pclk, *export_88, *gpio88_value, *gpio88_direction;
int mux_mode_88 = 0x8007, pin_number_88 = 88, logic_status_88 = 1;
char* pin_direction_88 = output;


//LCD_DB7: bone.P8_30: LCD Data line DB0: GPIO2_25: pin-89
FILE *ofp_lcd_ac_bias_en, *export_89, *gpio89_value, *gpio89_direction;
int mux_mode_89 = 0x8007, pin_number_89 = 89, logic_status_89 = 1;
char* pin_direction_89 = output;



int main(void)
{
configure_LCD_interface();
printf("Configure LCD \n");
LCD_init();  //call LCD initialize
printf("LCD initialize \n");

while(1)
 {
  LCD_print(1, "0123456789");
  printf("0123456789\n");
  delay_us(100);
  LCD_print(2, "Bonescript");
  printf("Bonescript\n");
  delay_us(100);
  }
return 1;
}


//*****************************************************************
//void configure_LCD_interface(void)
//*****************************************************************

void configure_LCD_interface(void)
{

//gpio44 mux setting
ofp_gpmc_ad12 = fopen("/sys/kernel/debug/omap_mux/gpmc_ad12", "w");
if(ofp_gpmc_ad12 == NULL) {printf("Unable to open gpmc_ad12.\n");}
```

```
fseek(ofp_gpmc_ad12, 0, SEEK_SET);
fprintf(ofp_gpmc_ad12, "0x%02x", mux_mode_44);
fflush(ofp_gpmc_ad12);

//create direction and value file for gpio44
export_44 = fopen("/sys/class/gpio/export", "w");
if(export_44 == NULL) {printf("Unable to open export.\n");}
fseek(export_44, 0, SEEK_SET);
fprintf(export_44, "%d", pin_number_44);
fflush(export_44);

//configure gpio44 direction
gpio44_direction = fopen("/sys/class/gpio/gpio44/direction", "w");
if(gpio44_direction == NULL) {printf("Unable to open gpio44_direction.\n");}
fseek(gpio44_direction, 0, SEEK_SET);
fprintf(gpio44_direction, "%s",  pin_direction_44);
fflush(gpio44_direction);
gpio44_value = fopen("/sys/class/gpio/gpio44/value", "w");
if(gpio44_value == NULL) {printf("Unable to open gpio44_value.\n");}


//LCD_E
//gpio26 mux setting
ofp_gpmc_ad10 = fopen("/sys/kernel/debug/omap_mux/gpmc_ad10", "w");
if(ofp_gpmc_ad10 == NULL) {printf("Unable to open gpmc_ad10.\n");}
fseek(ofp_gpmc_ad10, 0, SEEK_SET);
fprintf(ofp_gpmc_ad10, "0x%02x", mux_mode_26);
fflush(ofp_gpmc_ad10);

//create direction and value file for gpio26
export_26 = fopen("/sys/class/gpio/export", "w");
if(export_26 == NULL) {printf("Unable to open export.\n");}
fseek(export_26, 0, SEEK_SET);
fprintf(export_26, "%d", pin_number_26);
fflush(export_26);

//configure gpio26 direction
gpio26_direction = fopen("/sys/class/gpio/gpio26/direction", "w");
if(gpio26_direction == NULL) {printf("Unable to open gpio26_direction.\n");}
```

```
fseek(gpio26_direction, 0, SEEK_SET);
fprintf(gpio26_direction, "%s",  pin_direction_26);
fflush(gpio26_direction);
gpio26_value = fopen("/sys/class/gpio/gpio26/value", "w");
if(gpio26_value == NULL) {printf("Unable to open gpio26_value.\n");}


//LCD_DB0
//gpio46 mux setting
ofp_gpmc_ad14 = fopen("/sys/kernel/debug/omap_mux/gpmc_ad14", "w");
if(ofp_gpmc_ad14 == NULL) {printf("Unable to open gpmc_ad14.\n");}
fseek(ofp_gpmc_ad14, 0, SEEK_SET);
fprintf(ofp_gpmc_ad14, "0x%02x", mux_mode_46);
fflush(ofp_gpmc_ad14);

//create direction and value file for gpio46
export_46 = fopen("/sys/class/gpio/export", "w");
if(export_46 == NULL) {printf("Unable to open export.\n");}
fseek(export_46, 0, SEEK_SET);
fprintf(export_46, "%d", pin_number_46);
fflush(export_46);

//configure gpio46 direction
gpio46_direction = fopen("/sys/class/gpio/gpio46/direction", "w");
if(gpio46_direction == NULL) {printf("Unable to open gpio46_direction.\n");}
fseek(gpio46_direction, 0, SEEK_SET);
fprintf(gpio46_direction, "%s",  pin_direction_46);
fflush(gpio46_direction);
gpio46_value = fopen("/sys/class/gpio/gpio46/value", "w");
if(gpio46_value == NULL) {printf("Unable to open gpio46_value.\n");}


//LCD_DB1
//gpio65 mux setting
ofp_gpmc_clk = fopen("/sys/kernel/debug/omap_mux/gpmc_clk", "w");
if(ofp_gpmc_clk == NULL) {printf("Unable to open gpmc_clk.\n");}
fseek(ofp_gpmc_clk, 0, SEEK_SET);
fprintf(ofp_gpmc_clk, "0x%02x", mux_mode_65);
fflush(ofp_gpmc_clk);
```

```
//create direction and value file for gpio65
export_65 = fopen("/sys/class/gpio/export", "w");
if(export_65 == NULL) {printf("Unable to open export.\n");}
fseek(export_65, 0, SEEK_SET);
fprintf(export_65, "%d", pin_number_65);
fflush(export_65);


//configure gpio65 direction
gpio65_direction = fopen("/sys/class/gpio/gpio65/direction", "w");
if(gpio65_direction == NULL) {printf("Unable to open gpio65_direction.\n");}
fseek(gpio65_direction, 0, SEEK_SET);
fprintf(gpio65_direction, "%s", pin_direction_65);
fflush(gpio65_direction);
gpio65_value = fopen("/sys/class/gpio/gpio65/value", "w");
if(gpio65_value == NULL) {printf("Unable to open gpio65_value.\n");}



//LCD_DB2
//gpio63 mux setting
ofp_gpmc_csn2 = fopen("/sys/kernel/debug/omap_mux/gpmc_csn2", "w");
if(ofp_gpmc_csn2 == NULL) {printf("Unable to open gpmc_csn2.\n");}
fseek(ofp_gpmc_csn2, 0, SEEK_SET);
fprintf(ofp_gpmc_csn2, "0x%02x", mux_mode_63);
fflush(ofp_gpmc_csn2);


//create direction and value file for gpio63
export_63 = fopen("/sys/class/gpio/export", "w");
if(export_63 == NULL) {printf("Unable to open export.\n");}
fseek(export_63, 0, SEEK_SET);
fprintf(export_63, "%d", pin_number_63);
fflush(export_63);


//configure gpio63 direction
gpio63_direction = fopen("/sys/class/gpio/gpio63/direction", "w");
if(gpio63_direction == NULL) {printf("Unable to open gpio63_direction.\n");}
fseek(gpio63_direction, 0, SEEK_SET);
fprintf(gpio63_direction, "%s", pin_direction_63);
fflush(gpio63_direction);
```

```
gpio63_value = fopen("/sys/class/gpio/gpio63/value", "w");
if(gpio63_value == NULL) {printf("Unable to open gpio63_value.\n");}


//LCD_DB3
//gpio37 mux setting
ofp_gpmc_ad5 = fopen("/sys/kernel/debug/omap_mux/gpmc_ad5", "w");
if(ofp_gpmc_ad5 == NULL) {printf("Unable to open gpmc_ad5.\n");}
fseek(ofp_gpmc_ad5, 0, SEEK_SET);
fprintf(ofp_gpmc_ad5, "0x%02x", mux_mode_37);
fflush(ofp_gpmc_ad5);


//create direction and value file for gpio37
export_37 = fopen("/sys/class/gpio/export", "w");
if(export_37 == NULL) {printf("Unable to open export.\n");}
fseek(export_37, 0, SEEK_SET);
fprintf(export_37, "%d", pin_number_37);
fflush(export_37);


//configure gpio37 direction
gpio37_direction = fopen("/sys/class/gpio/gpio37/direction", "w");
if(gpio37_direction==NULL){printf("Unable to open gpio37_direction.\n");}
fseek(gpio37_direction, 0, SEEK_SET);
fprintf(gpio37_direction, "%s",  pin_direction_37);
fflush(gpio37_direction);
gpio37_value = fopen("/sys/class/gpio/gpio37/value", "w");
if(gpio37_value == NULL) {printf("Unable to open gpio37_value.\n");}


//LCD_DB4
//gpio33 mux setting
ofp_gpmc_ad1 = fopen("/sys/kernel/debug/omap_mux/gpmc_ad1", "w");
if(ofp_gpmc_ad1 == NULL) {printf("Unable to open gpmc_ad1.\n");}
fseek(ofp_gpmc_ad1, 0, SEEK_SET);
fprintf(ofp_gpmc_ad1, "0x%02x", mux_mode_33);
fflush(ofp_gpmc_ad1);


//create direction and value file for gpio33
export_33 = fopen("/sys/class/gpio/export", "w");
```

```
if(export_33 == NULL) {printf("Unable to open export.\n");}
fseek(export_33, 0, SEEK_SET);
fprintf(export_33, "%d", pin_number_33);
fflush(export_33);

//configure gpio33 direction
gpio33_direction = fopen("/sys/class/gpio/gpio33/direction", "w");
if(gpio33_direction == NULL) {printf("Unable to open gpio33_direction.\n");}
fseek(gpio33_direction, 0, SEEK_SET);
fprintf(gpio33_direction, "%s",  pin_direction_33);
fflush(gpio33_direction);
gpio33_value = fopen("/sys/class/gpio/gpio33/value", "w");
if(gpio33_value == NULL) {printf("Unable to open gpio33_value.\n");}


//LCD_DB5
//gpio61 mux setting
ofp_gpmc_csn0=fopen("/sys/kernel/debug/omap_mux/gpmc_csn0", "w");
if(ofp_gpmc_csn0==NULL){printf("Unable to open gpmc_csn0.\n");}
fseek(ofp_gpmc_csn0, 0, SEEK_SET);
fprintf(ofp_gpmc_csn0, "0x%02x", mux_mode_61);
fflush(ofp_gpmc_csn0);

//create direction and value file for gpio61
export_61 = fopen("/sys/class/gpio/export", "w");
if(export_61 == NULL) {printf("Unable to open export.\n");}
fseek(export_61, 0, SEEK_SET);
fprintf(export_61, "%d", pin_number_61);
fflush(export_61);

//configure gpio61 direction
gpio61_direction = fopen("/sys/class/gpio/gpio61/direction", "w");
if(gpio61_direction == NULL) {printf("Unable to open gpio61_direction.\n");}
fseek(gpio61_direction, 0, SEEK_SET);
fprintf(gpio61_direction, "%s",  pin_direction_61);
fflush(gpio61_direction);
gpio61_value = fopen("/sys/class/gpio/gpio61/value", "w");
if(gpio61_value == NULL) {printf("Unable to open gpio61_value.\n");}
```

```c
//LCD_DB6
//gpio88 mux setting
ofp_lcd_pclk = fopen("/sys/kernel/debug/omap_mux/lcd_pclk", "w");
if(ofp_lcd_pclk == NULL) {printf("Unable to open lcd_pclk.\n");}
fseek(ofp_lcd_pclk, 0, SEEK_SET);
fprintf(ofp_lcd_pclk, "0x%02x", mux_mode_88);
fflush(ofp_lcd_pclk);

//create direction and value file for gpio88
export_88 = fopen("/sys/class/gpio/export", "w");
if(export_88 == NULL) {printf("Unable to open export.\n");}
fseek(export_88, 0, SEEK_SET);
fprintf(export_88, "%d", pin_number_88);
fflush(export_88);

//configure gpio88 direction
gpio88_direction = fopen("/sys/class/gpio/gpio88/direction", "w");
if(gpio88_direction==NULL){printf("Unable to open gpio88_direction.\n");}
fseek(gpio88_direction, 0, SEEK_SET);
fprintf(gpio88_direction, "%s",  pin_direction_88);
fflush(gpio88_direction);
gpio88_value = fopen("/sys/class/gpio/gpio88/value", "w");
if(gpio88_value == NULL) {printf("Unable to open gpio88_value.\n");}


//LCD_DB7
//gpio89 mux setting
ofp_lcd_ac_bias_en=fopen("/sys/kernel/debug/omap_mux/lcd_ac_bias_en", "w");
if(ofp_lcd_ac_bias_en==NULL){printf("Unable to open lcd_ac_bias_en.\n");}
fseek(ofp_lcd_ac_bias_en, 0, SEEK_SET);
fprintf(ofp_lcd_ac_bias_en, "0x%02x", mux_mode_89);
fflush(ofp_lcd_ac_bias_en);

//create direction and value file for gpio89
export_89 = fopen("/sys/class/gpio/export", "w");
if(export_89 == NULL) {printf("Unable to open export.\n");}
fseek(export_89, 0, SEEK_SET);
fprintf(export_89, "%d", pin_number_89);
```

```
fflush(export_89);

//configure gpio89 direction
gpio89_direction = fopen("/sys/class/gpio/gpio89/direction", "w");
if(gpio89_direction == NULL) {printf("Unable to open gpio89_direction.\n");}
fseek(gpio89_direction, 0, SEEK_SET);
fprintf(gpio89_direction, "%s",  pin_direction_89);
fflush(gpio89_direction);
gpio89_value = fopen("/sys/class/gpio/gpio89/value", "w");
if(gpio89_value == NULL) {printf("Unable to open gpio89_value.\n");}
}


//**********************************************************************
//LCD_init
//**********************************************************************

void LCD_init(void)
{
delay_us(15000);                    //wait 15 ms
LCD_putcommand(0x30);               //set for 8-bit operation
delay_us(5000);                     //delay 5 ms
LCD_putcommand(0x30);               //set for 8-bit operation
delay_us(100);                      //delay 100 us
LCD_putcommand(0x30);               //set for 8-bit operation
LCD_putcommand(0x38);               //function set
LCD_putcommand(0x80);               //display off
LCD_putcommand(0x01);               //display clear
LCD_putcommand(0x06);               //entry mode set
}


//**********************************************************************
//LCD_putcommand
//**********************************************************************

void LCD_putcommand(unsigned char cmd)
{
//parse command variable into individual bits for output
//to LCD
```

```c
//configure DB7 value
fseek(gpio89_value, 0, SEEK_SET);
if((cmd & 0x0080)== 0x0080)
  {
  printf("CmDB7:1");
  logic_status_89 = 1;
  fprintf(gpio89_value, "%d", logic_status_89);
  }
else
  {
  printf("CmDB7:0");
  logic_status_89 = 0;
  fprintf(gpio89_value, "%d", logic_status_89);
  }
fflush(gpio89_value);


//configure DB6 value
fseek(gpio88_value, 0, SEEK_SET);
if((cmd & 0x0040)== 0x0040)
  {
  printf(" CmDB6:1");
  logic_status_88 = 1;
  fprintf(gpio88_value, "%d", logic_status_88);
  }
else
  {
  printf(" CmDB6:0");
  logic_status_88 = 0;
  fprintf(gpio88_value, "%d", logic_status_88);
  }
fflush(gpio88_value);


//configure DB5 value
fseek(gpio61_value, 0, SEEK_SET);
if((cmd & 0x0020)== 0x0020)
  {
  printf(" CmDB5:1");
```

```
  logic_status_61 = 1;
  fprintf(gpio61_value, "%d", logic_status_61);
  }
else
  {
  printf(" CmDB5:0");
  logic_status_61 = 0;
  fprintf(gpio61_value, "%d", logic_status_61);
  }
fflush(gpio61_value);


//configure DB4 value
fseek(gpio33_value, 0, SEEK_SET);
if((cmd & 0x0010)== 0x0010)
  {
  printf(" CmDB4:1");
  logic_status_33 = 1;
  fprintf(gpio33_value, "%d", logic_status_33);
  }
else
  {
  printf(" CmDB4:0");
  logic_status_33 = 0;
  fprintf(gpio33_value, "%d", logic_status_33);
  }
fflush(gpio33_value);


//configure DB3 value
fseek(gpio37_value, 0, SEEK_SET);
if((cmd & 0x0008)== 0x0008)
  {
  printf(" CmDB3:1");
  logic_status_37 = 1;
  fprintf(gpio37_value, "%d", logic_status_37);
  }
else
  {
```

```
  printf(" CmDB3:0");
  logic_status_37 = 0;
  fprintf(gpio37_value, "%d", logic_status_37);
  }
 fflush(gpio37_value);


 //configure DB2 value
 fseek(gpio63_value, 0, SEEK_SET);
 if((cmd & 0x0004)== 0x0004)
   {
   printf(" CmDB2:1");
   logic_status_63 = 1;
   fprintf(gpio63_value, "%d", logic_status_63);
   }
 else
   {
   printf(" CmDB2:0");
   logic_status_63 = 0;
   fprintf(gpio63_value, "%d", logic_status_63);
   }
 fflush(gpio63_value);


 //configure DB1 value
 fseek(gpio65_value, 0, SEEK_SET);
 if((cmd & 0x0002)== 0x0002)
   {
   printf(" CmDB1:1");
   logic_status_65 = 1;
   fprintf(gpio65_value, "%d", logic_status_65);
   }
 else
   {
   printf(" CmDB1:0");
   logic_status_65 = 0;
   fprintf(gpio65_value, "%d", logic_status_65);
   }
 fflush(gpio65_value);
```

```
//configure DB0 value
fseek(gpio46_value, 0, SEEK_SET);
if((cmd & 0x0001)== 0x0001)
  {
  printf(" CmDB0:1\n");
  logic_status_46 = 1;
  fprintf(gpio46_value, "%d", logic_status_46);
  }
else
  {
  printf(" CmDB0:0\n");
  logic_status_46 = 0;
  fprintf(gpio46_value, "%d", logic_status_46);
  }
fflush(gpio46_value);

//LCD Register Set (RS) to logic zero for command input
logic_status_44 = 0;
fseek(gpio44_value, 0, SEEK_SET);
fprintf(gpio44_value, "%d", logic_status_44);
fflush(gpio44_value);


//LCD Enable (E) pin high
logic_status_26 = 1;
fseek(gpio26_value, 0, SEEK_SET);
fprintf(gpio26_value, "%d", logic_status_26);
fflush(gpio26_value);

//delay
delay_us(2);

//LCD Enable (E) pin low
logic_status_26 = 0;
fseek(gpio26_value, 0, SEEK_SET);
fprintf(gpio26_value, "%d", logic_status_26);
fflush(gpio26_value);
```

```
//delay
delay_us(100);
}


//*****************************************************************
//LCD_putchar
//*****************************************************************

void LCD_putchar(unsigned char chr)
{
//parse character variable into individual bits for output
//to LCD

printf("Data: %c:%d", chr, chr);
chr = (int)(chr);

//configure DB7 value
fseek(gpio89_value, 0, SEEK_SET);
if((chr & 0x0080)== 0x0080)
   {
   printf(" DB7:1");
   logic_status_89 = 1;
   fprintf(gpio89_value, "%d", logic_status_89);
   }
else
   {
   printf(" DB7:0");
   logic_status_89 = 0;
   fprintf(gpio89_value, "%d", logic_status_89);
   }
fflush(gpio89_value);


//configure DB6 value
fseek(gpio88_value, 0, SEEK_SET);
if((chr & 0x0040)== 0x0040)
   {
   printf(" DB6:1");
```

```
  logic_status_88 = 1;
  fprintf(gpio88_value, "%d", logic_status_88);
  }
else
  {
  printf(" DB6:0");
  logic_status_88 = 0;
  fprintf(gpio88_value, "%d", logic_status_88);
  }
fflush(gpio88_value);


//configure DB5 value
fseek(gpio61_value, 0, SEEK_SET);
if((chr & 0x0020)== 0x0020)
  {
  printf(" DB5:1");
  logic_status_61 = 1;
  fprintf(gpio61_value, "%d", logic_status_61);
  }
else
  {
  printf(" DB5:0");
  logic_status_61 = 0;
  fprintf(gpio61_value, "%d", logic_status_61);
  }
fflush(gpio61_value);


//configure DB4 value
fseek(gpio33_value, 0, SEEK_SET);
if((chr & 0x0010)== 0x0010)
  {
  printf(" DB4:1");
  logic_status_33 = 1;
  fprintf(gpio33_value, "%d", logic_status_33);
  }
else
  {
```

```
   printf(" DB4:0");
   logic_status_33 = 0;
   fprintf(gpio33_value, "%d", logic_status_33);
   }
fflush(gpio33_value);


//configure DB3 value
fseek(gpio37_value, 0, SEEK_SET);
if((chr & 0x0008)== 0x0008)
   {
   printf(" DB3:1");
   logic_status_37 = 1;
   fprintf(gpio37_value, "%d", logic_status_37);
   }
else
   {
   printf(" DB3:0");
   logic_status_37 = 0;
   fprintf(gpio37_value, "%d", logic_status_37);
   }
fflush(gpio37_value);


//configure DB2 value
fseek(gpio63_value, 0, SEEK_SET);
if((chr & 0x0004)== 0x0004)
   {
   printf(" DB2:1");
   logic_status_63 = 1;
   fprintf(gpio63_value, "%d", logic_status_63);
   }
else
   {
   printf(" DB2:0");
   logic_status_63 = 0;
   fprintf(gpio63_value, "%d", logic_status_63);
   }
fflush(gpio63_value);
```

```
//configure DB1 value
fseek(gpio65_value, 0, SEEK_SET);
if((chr & 0x0002)== 0x0002)
  {
  printf(" DB1:1");
  logic_status_65 = 1;
  fprintf(gpio65_value, "%d", logic_status_65);
  }
else
  {
  printf(" DB1:0");
  logic_status_65 = 0;
  fprintf(gpio65_value, "%d", logic_status_65);
  }
fflush(gpio65_value);

//configure DB0 value
fseek(gpio46_value, 0, SEEK_SET);
if((chr & 0x0001)== 0x0001)
  {
  printf(" DB0:1\n");
  logic_status_46 = 1;
  fprintf(gpio46_value, "%d", logic_status_46);
  }
else
  {
  printf(" DB0:0\n");
  logic_status_46 = 0;
  fprintf(gpio46_value, "%d", logic_status_46);
  }
fflush(gpio46_value);

//LCD Register Set (RS) to logic one for character input
logic_status_44 = 1;
fseek(gpio44_value, 0, SEEK_SET);
fprintf(gpio44_value, "%d", logic_status_44);
fflush(gpio44_value);
```

```c
//LCD Enable (E) pin high
logic_status_26 = 1;
fseek(gpio26_value, 0, SEEK_SET);
fprintf(gpio26_value, "%d", logic_status_26);
fflush(gpio26_value);

//delay
delay_us(2);

//LCD Enable (E) pin low
logic_status_26 = 0;
fseek(gpio26_value, 0, SEEK_SET);
fprintf(gpio26_value, "%d", logic_status_26);
fflush(gpio26_value);

//delay
delay_us(2);

}

//*******************************************************************
//*******************************************************************
void LCD_print(unsigned int line, char *msg)
{
int i = 0;

if(line == 1)
  {
  LCD_putcommand(0x80);            //print to LCD line 1
  }
else
  {
  LCD_putcommand(0xc0);            //print to LCD line 2
  }

while(*(msg) != '\0')
  {
```

```
  LCD_putchar(*msg);
  //printf("Data: %c\n\n", *msg);
  msg++;
  }
}


//*****************************************************************

void delay_us(int desired_delay_us)
{
struct timeval  tv_start;  //start time hack
struct timeval  tv_now;    //current time hack
int elapsed_time_us;

gettimeofday(&tv_start, NULL);
elapsed_time_us = 0;

while(elapsed_time_us <  desired_delay_us)
  {
  gettimeofday(&tv_now, NULL);
  if(tv_now.tv_usec >= tv_start.tv_usec)
    elapsed_time_us = tv_now.tv_usec - tv_start.tv_usec;
  else
    elapsed_time_us = (1000000 - tv_start.tv_usec) + tv_now.tv_usec;
  //printf("start: %ld \n", tv_start.tv_usec);
  //printf("now: %ld \n", tv_now.tv_usec);
  //printf("desired: %d \n", desired_delay_ms);
  //printf("elapsed: %d \n\n", elapsed_time_ms);
  }
}


//*****************************************************************
```

B.2   BEAGLEBONE BLACK –LINUX 3.8

Provided below is the C code for the BeagleBone Black operating under Linux–3.8.

   Please note the following pins were used in the interface:

- P8.9–RS

- P8.10–E

- P8.11–DB0

- P8.12–DB1

- P8.13–DB2

- P8.14–DB3

- P8.15–DB4

- P8.16–DB5

- P8.17–DB6

- P8.18–DB7

```c
//***************************************************************
#include <stdio.h>
#include <stdlib.h>
#include <stddef.h>
#include <time.h>
#include <string.h>
#include <sys/time.h>

#define  output "out"
#define  input  "in"

//function prototypes
void configure_LCD_interface(void);
void LCD_putcommand(unsigned char);
void LCD_putchar(unsigned char);
void LCD_print(unsigned int, char *string);
void delay_us(int);
void LCD_init(void);
FILE *setup_gpio(unsigned int pin_number, char* direction);
void set_gpio_value(FILE* value_file, unsigned int logic_status);

//LCD_RS: bone.P8_9 LCD Register Set (RS) control: GPIO2_5: pin-69
FILE* gpio_rs_value = NULL;
int pin_number_rs = 69, logic_status_rs = 1;
char* pin_direction_rs = output;

//LCD_E:   bone.P8_10: LCD Enable En (E): GPIO2_4: pin-68
```

```
FILE* gpio_e_value = NULL;
int pin_number_e = 68, logic_status_e = 1;
char* pin_direction_e = output;

//LCD_DB0: bone.P8_11: LCD Data line DB0: GPIO1_13: pin-45
FILE *gpio_db0_value = NULL;
int pin_number_db0 = 45, logic_status_db0 = 1;
char* pin_direction_db0 = output;

//LCD_DB1: bone.P8_12: LCD Data line DB1: GPIO1_12: pin-44
FILE *gpio_db1_value = NULL;
int pin_number_db1 = 44, logic_status_db1 = 1;
char* pin_direction_db1 = output;

//LCD_DB2: bone.P8_13: LCD Data line DB2: GPIO0_23: pin-23
FILE *gpio_db2_value = NULL;
int pin_number_db2 = 23, logic_status_db2 = 1;
char* pin_direction_db2 = output;

//LCD_DB3: bone.P8_14: LCD Data line DB3: GPIO0_26: pin-26
FILE *gpio_db3_value = NULL;
int pin_number_db3 = 26, logic_status_db3 = 1;
char* pin_direction_db3 = output;

//LCD_DB4: bone.P8_15: LCD Data line DB4: GPIO1_15: pin-47
FILE *gpio_db4_value = NULL;
int pin_number_db4 = 47, logic_status_db4 = 1;
char* pin_direction_db4 = output;

//LCD_DB5: bone.P8_16: LCD Data line DB5: GPIO1_14: pin-46
FILE *gpio_db5_value = NULL;
int pin_number_db5 = 46, logic_status_db5 = 1;
char* pin_direction_db5 = output;

//LCD_DB6: bone.P8_17: LCD Data line DB6: GPIO0_27: pin-27
FILE *gpio_db6_value = NULL;
int pin_number_db6 = 27, logic_status_db6 = 1;
char* pin_direction_db6 = output;
```

```
//LCD_DB7: bone.P8_18: LCD Data line DB7: GPIO2_1: pin-65
FILE *gpio_db7_value = NULL;
int pin_number_db7 = 65, logic_status_db7 = 1;
char* pin_direction_db7 = output;


int main(void)
{
configure_LCD_interface();
printf("Configure LCD \n");
LCD_init();  //call LCD initialize
printf("LCD initialize \n");

while(1)
 {
  LCD_print(1, "Bad to the");
  printf("Bad to the\n");
  delay_us(100);
  LCD_print(2, "  Bone");
  printf("  Bone\n");
  delay_us(100);
  }
return 1;
}


//********************************************************************
//FILE *setup_gpio(unsigned int pin_number, char* pin_direction)
//********************************************************************

FILE *setup_gpio(unsigned int pin_number, char* pin_direction)
{
FILE *exp, *gpio_value, *gpio_direction;
char gpio_direction_filename[40];
char gpio_value_filename[40];

//create direction and value file for pin
exp = fopen("/sys/class/gpio/export", "w");
if(exp == NULL) {printf("Unable to open export.\n");}
fseek(exp, 0, SEEK_SET);
```

```
fprintf(exp, "%d", pin_number);
fflush(exp);
fclose(exp);

//configure pin direction
sprintf(gpio_direction_filename, "/sys/class/gpio/gpio%d/direction",
        pin_number);
gpio_direction = fopen(gpio_direction_filename, "w");
if(gpio_direction == NULL) {printf("Unable to open %s.\n",
                                   gpio_direction_filename);}
fseek(gpio_direction, 0, SEEK_SET);
fprintf(gpio_direction, "%s", pin_direction);
fflush(gpio_direction);
fclose(gpio_direction);

//open pin value file
sprintf(gpio_value_filename, "/sys/class/gpio/gpio%d/value", pin_number);
gpio_value = fopen(gpio_value_filename, "w");
if(gpio_value==NULL){printf("Unable to open %s.\n", gpio_value_filename);}

printf("Opening %d (%s) returned 0x%08x\n", pin_number, gpio_value_filename,
        (int)gpio_value);
return(gpio_value);
}

//*******************************************************************
//void set_gpio_value(FILE* value_file, unsigned int logic_status)
//*******************************************************************
void set_gpio_value(FILE* value_file, unsigned int logic_status)
{

printf(" %d -> 0x%08x\n", logic_status, (int)value_file);
fseek(value_file, 0, SEEK_SET);
fprintf(value_file, "%d", logic_status);
fflush(value_file);

}

//*******************************************************************
```

```c
//void configure_LCD_interface(void)
//********************************************************************


void configure_LCD_interface(void)
{

//Setup LCD GPIO pins
gpio_rs_value = setup_gpio(pin_number_rs, pin_direction_rs);
gpio_e_value = setup_gpio(pin_number_e, pin_direction_e);
gpio_db0_value = setup_gpio(pin_number_db0, pin_direction_db0);
gpio_db1_value = setup_gpio(pin_number_db1, pin_direction_db1);
gpio_db2_value = setup_gpio(pin_number_db2, pin_direction_db2);
gpio_db3_value = setup_gpio(pin_number_db3, pin_direction_db3);
gpio_db4_value = setup_gpio(pin_number_db4, pin_direction_db4);
gpio_db5_value = setup_gpio(pin_number_db5, pin_direction_db5);
gpio_db6_value = setup_gpio(pin_number_db6, pin_direction_db6);
gpio_db7_value = setup_gpio(pin_number_db7, pin_direction_db7);

}


//********************************************************************
//LCD_init
//********************************************************************


void LCD_init(void)
{
delay_us(15000);                     //wait 15 ms
LCD_putcommand(0x30);                //set for 8-bit operation
delay_us(5000);                      //delay 5 ms
LCD_putcommand(0x30);                //set for 8-bit operation
delay_us(100);                       //delay 100 us
LCD_putcommand(0x30);                //set for 8-bit operation
LCD_putcommand(0x38);                //function set
LCD_putcommand(0x80);                //display off
LCD_putcommand(0x01);                //display clear
LCD_putcommand(0x06);                //entry mode set
}


//********************************************************************
```

```
//LCD_putcommand
//*********************************************************************

void LCD_putcommand(unsigned char cmd)
{
//parse command variable into individual bits for output
//to LCD

//configure DB7 value
if((cmd & 0x0080)== 0x0080)
  {
  printf("CmDB7:1");
  logic_status_db7 = 1;
  }
else
  {
  printf(" CmDB7:0");
  logic_status_db7 = 0;
  }
set_gpio_value(gpio_db7_value, logic_status_db7);


//configure DB6 value
if((cmd & 0x0040)== 0x0040)
  {
  printf(" CmDB6:1");
  logic_status_db6 = 1;
  }
else
  {
  printf(" CmDB6:0");
  logic_status_db6 = 0;
  }
set_gpio_value(gpio_db6_value, logic_status_db6);


//configure DB5 value
if((cmd & 0x0020)== 0x0020)
  {
```

```
  printf(" CmDB5:1");
  logic_status_db5 = 1;
  }
else
  {
  printf(" CmDB5:0");
  logic_status_db5 = 0;
  }
set_gpio_value(gpio_db5_value, logic_status_db5);


//configure DB4 value
if((cmd & 0x0010)== 0x0010)
  {
  printf(" CmDB4:1");
  logic_status_db4 = 1;
  }
else
  {
  printf(" CmDB4:0");
  logic_status_db4 = 0;
  }
set_gpio_value(gpio_db4_value, logic_status_db4);


//configure DB3 value
if((cmd & 0x0008)== 0x0008)
  {
  printf(" CmDB3:1");
  logic_status_db3 = 1;
  }
else
  {
  printf(" CmDB3:0");
  logic_status_db3 = 0;
  }
set_gpio_value(gpio_db3_value, logic_status_db3);
```

```
//configure DB2 value
if((cmd & 0x0004)== 0x0004)
  {
  printf(" CmDB2:1");
  logic_status_db2 = 1;
  }
else
  {
  printf(" CmDB2:0");
  logic_status_db2 = 0;
  }
set_gpio_value(gpio_db2_value, logic_status_db2);


//configure DB1 value
if((cmd & 0x0002)== 0x0002)
  {
  printf(" CmDB1:1");
  logic_status_db1 = 1;
  }
else
  {
  printf(" CmDB1:0");
  logic_status_db1 = 0;
  }
set_gpio_value(gpio_db1_value, logic_status_db1);


//configure DB0 value
if((cmd & 0x0001)== 0x0001)
  {
  printf(" CmDB0:1");
  logic_status_db0 = 1;
  }
else
  {
  printf(" CmDB0:0");
  logic_status_db0 = 0;
  }
```

```
set_gpio_value(gpio_db0_value, logic_status_db0);

printf("\n");

//LCD Register Set (RS) to logic zero for command input
logic_status_rs = 0;
set_gpio_value(gpio_rs_value, logic_status_rs);

//LCD Enable (E) pin high
logic_status_e = 1;
set_gpio_value(gpio_e_value, logic_status_e);

//delay
delay_us(2);

//LCD Enable (E) pin low
logic_status_e = 0;
set_gpio_value(gpio_e_value, logic_status_e);

//delay
delay_us(100);
}

//********************************************************************
//LCD_putchar
//********************************************************************

void LCD_putchar(unsigned char chr)
{
//parse character variable into individual bits for output
//to LCD

printf("Data: %c:%d", chr, chr);
chr = (int)(chr);

//configure DB7 value
if((chr & 0x0080)== 0x0080)
  {
  printf(" DB7:1");
```

```
  logic_status_db7 = 1;
  }
else
  {
  printf(" DB7:0");
  logic_status_db7 = 0;
  }
set_gpio_value(gpio_db7_value, logic_status_db7);


//configure DB6 value
if((chr & 0x0040)== 0x0040)
  {
  printf(" DB6:1");
  logic_status_db6 = 1;
  }
else
  {
  printf(" DB6:0");
  logic_status_db6 = 0;
  }
set_gpio_value(gpio_db6_value, logic_status_db6);


//configure DB5 value
if((chr & 0x0020)== 0x0020)
  {
  printf(" DB5:1");
  logic_status_db5 = 1;
  }
else
  {
  printf(" DB5:0");
  logic_status_db5 = 0;
  }
set_gpio_value(gpio_db5_value, logic_status_db5);


//configure DB4 value
```

```
if((chr & 0x0010)== 0x0010)
  {
  printf(" DB4:1");
  logic_status_db4 = 1;
  }
else
  {
  printf(" DB4:0");
  logic_status_db4 = 0;
  }
set_gpio_value(gpio_db4_value, logic_status_db4);


//configure DB3 value
if((chr & 0x0008)== 0x0008)
  {
  printf(" DB3:1");
  logic_status_db3 = 1;
  }
else
  {
  printf(" DB3:0");
  logic_status_db3 = 0;
  }
set_gpio_value(gpio_db3_value, logic_status_db3);


//configure DB2 value
if((chr & 0x0004)== 0x0004)
  {
  printf(" DB2:1");
  logic_status_db2 = 1;
  }
else
  {
  printf(" DB2:0");
  logic_status_db2 = 0;
  }
set_gpio_value(gpio_db2_value, logic_status_db2);
```

```
//configure DB1 value
if((chr & 0x0002)== 0x0002)
  {
  printf(" DB1:1");
  logic_status_db1 = 1;
  }
else
  {
  printf(" DB1:0");
  logic_status_db1 = 0;
  }
set_gpio_value(gpio_db1_value, logic_status_db1);

//configure DB0 value
if((chr & 0x0001)== 0x0001)
  {
  printf(" DB0:1\n");
  logic_status_db0 = 1;
  }
else
  {
  printf(" DB0:0\n");
  logic_status_db0 = 0;
  }
set_gpio_value(gpio_db0_value, logic_status_db0);

//LCD Register Set (RS) to logic one for character input
logic_status_rs = 1;
set_gpio_value(gpio_rs_value, logic_status_rs);

//LCD Enable (E) pin high
logic_status_e = 1;
set_gpio_value(gpio_e_value, logic_status_e);

//delay
delay_us(2);
```

```c
//LCD Enable (E) pin low
logic_status_e = 0;
set_gpio_value(gpio_e_value, logic_status_e);

//delay
delay_us(2);

}

//*******************************************************************
//*******************************************************************
void LCD_print(unsigned int line, char *msg)
{
int i = 0;

if(line == 1)
  {
  LCD_putcommand(0x80);           //print to LCD line 1
  }
else
  {
  LCD_putcommand(0xc0);           //print to LCD line 2
  }

while(*(msg) != '\0')
  {
  LCD_putchar(*msg);
  //printf("Data: %c\n\n", *msg);
  msg++;
  }
}


//*******************************************************************

void delay_us(int desired_delay_us)
{
struct timeval  tv_start;  //start time hack
struct timeval  tv_now;    //current time hack
int elapsed_time_us;
```

```
gettimeofday(&tv_start, NULL);
elapsed_time_us = 0;

while(elapsed_time_us <  desired_delay_us)
  {
  gettimeofday(&tv_now, NULL);
  if(tv_now.tv_usec >= tv_start.tv_usec)
    elapsed_time_us = tv_now.tv_usec - tv_start.tv_usec;
  else
    elapsed_time_us = (1000000 - tv_start.tv_usec) + tv_now.tv_usec;
  //printf("start: %ld \n", tv_start.tv_usec);
  //printf("now: %ld \n", tv_now.tv_usec);
  //printf("desired: %d \n", desired_delay_ms);
  //printf("elapsed: %d \n\n", elapsed_time_ms);
  }
}


//*****************************************************************
```

A P P E N D I X C

Parts List for Projects

Chapter 1			
Description	**Qty**	**Source**	**Part number**
5 VDC, 2A power supply	1	Adafruit (www.adafruit.com)	276
LED, red	1	Jameco (www.jameco.com)	333973
220 ohm resistor, 1/4W	1	Jameco (www.jameco.com)	690700
Boneyard I			
black pelican micro case	1	Pelican Cases/www.pelican.com	1040
3.3 x 2.1 solderless breadboad	1	Jameco/www.jameco.com	226094
circuit board hardware mounting hardware	1	Jameco (www.jameco.com)	106551

Chapter 2			
Description	**Qty**	**Source**	**Part number**
Blinky 602A robot	1	Graymark(www.graymarkint.com)	Blinky 602A
aluminum bracket	1	local manufacture	n/a
bracket HW			
- screws	9	Jameco (www.jameco.com)	40970
- nuts	9	Jameco (www.jameco.com)	40943
- washer	9	Jameco (www.jameco.com)	106286
Sharp IR sensor GP2D120XJ00F	3	Sparkfun (www.sparkfun.com)	SEN-08959
sensor cables	3	Sparkfun (www.sparkfun.com)	SEN-08733
10k ohm resistor, 1/4W	1	Jameco (www.jameco.com)	691104
2N2222 transistor	1	Jameco (www.jameco.com)	38236
LED, red	1	Jameco (www.jameco.com)	333973
220 ohm resistor, 1/4W	1	Jameco (www.jameco.com)	690700
1M ohm trim potentiometer	3	Jameco (www.jameco.com)	254749
white LED	1	Jameco (www.jameco.com)	664341
TIP 120 Darlington NPN transistor	2	Jameco (www.jameco.com)	803671
1N4001 diode	2	Jameco (www.jameco.com)	35975
240 ohm resistor	2	Jameco (www.jameco.com)	690718
7805 5 VDC, 1A voltage regulator	1	Jameco (www.jameco.com)	51262
9 VDC, 2A power supply	1	Jameco (www.jameco.com)	1952847
LM1084-3.3, 3.3 VDC voltage regulator	1	Jameco (www.jameco.com)	299735

Chapter 3			
Description	**Qty**	**Source**	**Part number**
LCD7 display	1	circuitco (www.circuitco.com)	LCD7
Prototype Cape Kit for BeagleBone	1	Adafruit (www.adafruit.com)	572
prototype builder, 1.6" x 2.7"	1	Jameco (www.jameco.com)	105100
circuit board hardware mounting kit	1	Jameco (www.jameco.com)	106551

Figure C.1: Parts list.

Chapter 4

Description	Qty	Source	Part number
LM117T voltage regulator	1	Jameco (www.jameco.com)	23579
0.1 uF, 25 VDC capacitor	1	Jameco (www.jameco.com)	151116
5K ohm trim potentiometer	1	Jameco (www.jameco.com)	254669
240 ohm resistor, 1/4W	1	Jameco (www.jameco.com)	690718
1 uf, 25 VDC capacitor	1	Jameco (www.jameco.com)	330431
Prototype Cape kit for BeagleBone	1	Adafruit (www.adafruit.com)	572
thumb joystick	1	Sparkfun (www.sparkfun.com)	COM-09032
Breakout board for thumb joystick	1	Sparkfun (www.sparkfun.com)	BOB-09110
10 ohm resistor, 1/4W	1	Jameco (www.jameco.com)	690380
100 VDC, 5A Schottky diode (IR 50SQ100)	1	Digikey (www.digikey.com)	50SQ100CT-ND
thrusters, Shoreline Bilge Pump	3	Walmart (www.walmart.com)	user choice
Sharp IR sensor GP2D120XJ00F	3	Sparkfun (www.sparkfun.com)	SEN-08959
1M ohm trim pots	3	Jameco (www.jameco.com)	254749
LED, red	1	Jameco (www.jameco.com)	333973
3.3 VDC to 5 VDC level shifter (T.I. PCA 9306)	1	Jameco (www.jameco.com)	296-17988-1-ND
7404 hex inverter	1	Jameco (www.jameco.com)	49040
7408 quad AND gate	1	Jameco (www.jameco.com)	49146
200 ohm resistor, 1/4W	4	Jameco (www.jameco.com)	690697
TIP 31 NPN transistor	8	Jameco (www.jameco.com)	179354
TIP 32 PNP transistor	4	Jameco (www.jameco.com)	181841
100 VDC, 5A Schottky diode (IR 50SQ100)	8	Jameco (www.jameco.com)	50SQ100CT-ND
470 ohm resistor, 1/4W	4	Jameco (www.jameco.com)	690785
1000 uF, 25 VDC capacitor	2	Jameco (www.jameco.com)	158298
4WD robot platform (DF ROBOT, ROB0003)	1	Jameco (www.jameco.com)	2124285

Chapter 5

Description	Qty	Source	Part number
LED, red	1	Jameco (www.jameco.com)	333973
LED, green	1	Jameco (www.jameco.com)	34761
220 ohm resistor, 1/4W	1	Jameco (www.jameco.com)	690700
4.7K ohm resistor, 1/4W	1	Jameco (www.jameco.com)	691024
tact pushbutton switch	1	Jameco (www.jameco.com)	199726

Figure C.2: Parts list (continued).

Chapter 6			
Description	**Qty**	**Source**	**Part number**
Boneyard II			
LCD7	1	circuitco (www.circuitco.com)	LCD7
BeagleBone	1	multiple sources	---
Mini USB mouse	1	local purchase	---
USB hub	1	local purchase	---
Miniature keyboard	1	Adafruit (www.adafruit.com)	857
Pelican case	1	Pelican cases (www.pelican.com)	1200
Weather Station			
LM34DZ precision Fahrenheit temp sensor	1	Jameco (www.jameco.com)	155192
weather vane	1	multiple sources	---
LCD, 3.3 VDC, 16 x 2 char	1	Sparkfun (www.sparkfun.com)	LCD-09052
120 ohm resistor, 1/4W	1	Jameco (www.jameco.com)	690646
LED, red	8	Jameco (www.jameco.com)	333973
10K ohm resistor, 1/4W	8	Jameco (www.jameco.com)	691104
MPQ2222, general purpose NPN	8	Jameco (www.jameco.com)	26446
Speak and Spell			
text to speech chip	1	www.speechchips.com	SPO-512
mini-speaker, 8 ohm	1	Radioshack (www.radioshack.com)	273-0092
10K ohn resistor, 1/4W	2	Jameco (www.jameco.com)	691104
10 uF, 25 VDC	2	Jameco (www.jameco.com)	94212
4.7 uF, 25 VDC	1	Jameco (www.jameco.com)	2143460
330 ohm resistor, 1/4W	2	Jameco (www.jameco.com)	690742
LED, red	1	Jameco (www.jameco.com)	333973
LED, green	1	Jameco (www.jameco.com)	34761
10K ohm trim pot	1	Jameco (www.jameco.com)	254677
LM380N-3 2.5W audio amplifier	1	Jameco (www.jameco.com)	24037
0.1 uF, 25 VDC	1	Jameco (www.jameco.com)	151116
100 uF, 25 VDC	1	Jameco (www.jameco.com)	93761
Dagu 5 ROV			
Dagu Rover 5 tracked chassis	1	Jameco (www.jameco.com)	2143865
7.0 VDC, 1.5A power supply	2	Jameco (www.jameco.com)	249017
1N4001 silicon diode	2	Jameco (www.jameco.com)	35975
TIP 120, NPN Darlington	2	Jameco (www.jameco.com)	32993
330 ohm resistor	3	Jameco (www.jameco.com)	690742
Sharp IR sensor GP2D120XJ00F	3	Jameco (www.jameco.com)	SEN-08959
sensor cables	3	Jameco (www.jameco.com)	SEN-08733
1M ohm trim pot	1	Jameco (www.jameco.com)	254749
7805 5 VDC, 1A voltage regulator	1	Jameco (www.jameco.com)	51262
LM1084-3.3, 3.3 VDC voltage regulator	3	Jameco (www.jameco.com)	299735
100 uF, 25 VDC	1	Jameco (www.jameco.com)	93761
Stache Cam			
camera, PS3 Eye	1	multiple sources	PS3 Eye
LCD3	1	circuitco (www.circuitco.com)	LCD3
BeagleBone	1	multiple sources	---
Battery cape	1	circuitco (www.circuitco.com)	---

Figure C.3: Parts list (continued).

APPENDIX D

BeagleBone Device Tree

This Appendix contains the following files and examples:

- am33xx.dtsi: The devicetree data of the base processor.

- am335x–bone–common.dtsi: The devicetree data common for both BeagleBone and Beagle-Bone Black (includes am33xx.dtsi).

- am335x–bone.dts: The devicetree data for BeagleBone (includes am335x–bone–common.dtsi).

- am335x–boneblack.dts: The devicetree data for BeagleBone Black (includes am335x–bone–common.dtsi).

- am33xx_pwm–00A0.dts: A devicetree overlay that enables all of the PWMs on the am335x that can be accessed on the BeagleBone headers.

- bone_pwm_P8_13–00A0.dts: A devicetree overlay that configures the pinmux and PWM for BeagleBone header pin P8_13. This code example requires that am33xx_pwm be loaded first to enable the PWM subsystems before this overlay configures an individual PWM.

- cape–bone–iio–00A0.dts: A devicetree overlay that utilizes the bone–iio–helper to provide SYSFS entries for the ADCs.

D.1 AM33XX.DTSI

This section contains the file am33xx.dtsi which is the devicetree data of the base processor.

```
//******************************************************************
/*
 * Device Tree Source for AM33XX SoC
 *
 * Copyright (C) 2012 Texas Instruments Incorporated - http://www.ti.com/
 *
 * This file is licensed under the terms of the GNU General Public License
 * version 2.  This program is licensed "as is" without any warranty of any
 * kind, whether express or implied.
 */
```

```
/include/ "skeleton.dtsi"

/ {
    compatible = "ti,am33xx";
    interrupt-parent = <&intc>;

    aliases {
        serial0 = &uart1;
        serial1 = &uart2;
        serial2 = &uart3;
        serial3 = &uart4;
        serial4 = &uart5;
        serial5 = &uart6;
    };

    cpus {
        cpu: cpu@0 {
            compatible = "arm,cortex-a8";

            /*
             * To consider voltage drop between PMIC and SoC,
             * tolerance value is reduced to 2% from 4% and
             * voltage value is increased as a precaution.
             */
            operating-points = <
                /* kHz     uV */
                /* ES 2.0 Nitro and Turbo OPPs"
                1000000 1350000
                800000  1300000
                */
                720000  1285000
                600000  1225000
                500000  1125000
                275000  1125000
            >;
            voltage-tolerance = <2>; /* 2 percentage */
            clock-latency = <300000>; /* From omap-cpufreq driver */
        };
```

```
    };

    /*
     * The soc node represents the soc top level view.
It is uses for IPs
     * that are not memory mapped in the MPU view or for the MPU itself.
     */
    soc {
        compatible = "ti,omap-infra";
        mpu {
            compatible = "ti,omap3-mpu";
            ti,hwmods = "mpu";
        };
    };

    am33xx_pinmux: pinmux@44e10800 {
        compatible = "pinctrl-single";
        reg = <0x44e10800 0x0238>;
        #address-cells = <1>;
        #size-cells = <0>;
        pinctrl-single,register-width = <32>;
        pinctrl-single,function-mask = <0x7f>;
    };

    /*
     * XXX: Use a flat representation of the AM33XX interconnect.
     * The real AM33XX interconnect network is quite complex. Since
     * that will not bring real advantage to represent that in DT
     * for the moment, just use a fake OCP bus entry to represent
     * the whole bus hierarchy.
     */
    ocp {
        compatible = "simple-bus";
        #address-cells = <1>;
        #size-cells = <1>;
        ranges;
        ti,hwmods = "l3_main";

        intc: interrupt-controller@48200000 {
```

```
        compatible = "ti,omap2-intc";
        interrupt-controller;
        #interrupt-cells = <1>;
        ti,intc-size = <128>;
        reg = <0x48200000 0x1000>;
    };

    edma: edma@49000000 {
        compatible = "ti,edma3";
        ti,hwmods = "tpcc", "tptc0", "tptc1", "tptc2";
        reg =    <0x49000000 0x10000>,
            <0x44e10f90 0x10>;
        interrupt-parent = <&intc>;
        interrupts = <12 13 14>;
        #dma-cells = <1>;
        dma-channels = <64>;
        ti,edma-regions = <4>;
        ti,edma-slots = <256>;
        ti,edma-queue-tc-map = <0 0
                    1 1
                    2 2>;
        ti,edma-queue-priority-map = <0 0
                        1 1
                        2 2>;
        ti,edma-default-queue = <0>;
    };

    gpio1: gpio@44e07000 {
        compatible = "ti,omap4-gpio";
        ti,hwmods = "gpio1";
        gpio-controller;
        #gpio-cells = <2>;
        interrupt-controller;
        #interrupt-cells = <1>;
        reg = <0x44e07000 0x1000>;
        interrupts = <96>;
    };

    gpio2: gpio@4804c000 {
```

```
        compatible = "ti,omap4-gpio";
        ti,hwmods = "gpio2";
        gpio-controller;
        #gpio-cells = <2>;
        interrupt-controller;
        #interrupt-cells = <1>;
        reg = <0x4804c000 0x1000>;
        interrupts = <98>;
    };

    gpio3: gpio@481ac000 {
        compatible = "ti,omap4-gpio";
        ti,hwmods = "gpio3";
        gpio-controller;
        #gpio-cells = <2>;
        interrupt-controller;
        #interrupt-cells = <1>;
        reg = <0x481ac000 0x1000>;
        interrupts = <32>;
    };

    gpio4: gpio@481ae000 {
        compatible = "ti,omap4-gpio";
        ti,hwmods = "gpio4";
        gpio-controller;
        #gpio-cells = <2>;
        interrupt-controller;
        #interrupt-cells = <1>;
        reg = <0x481ae000 0x1000>;
        interrupts = <62>;
    };

    uart1: serial@44e09000 {
        compatible = "ti,omap3-uart";
        ti,hwmods = "uart1";
        clock-frequency = <48000000>;
        reg = <0x44e09000 0x2000>;
        interrupts = <72>;
        status = "disabled";
```

```
    };

    uart2: serial@48022000 {
        compatible = "ti,omap3-uart";
        ti,hwmods = "uart2";
        clock-frequency = <48000000>;
        reg = <0x48022000 0x2000>;
        interrupts = <73>;
        status = "disabled";
    };

    uart3: serial@48024000 {
        compatible = "ti,omap3-uart";
        ti,hwmods = "uart3";
        clock-frequency = <48000000>;
        reg = <0x48024000 0x2000>;
        interrupts = <74>;
        status = "disabled";
    };

    uart4: serial@481a6000 {
        compatible = "ti,omap3-uart";
        ti,hwmods = "uart4";
        clock-frequency = <48000000>;
        reg = <0x481a6000 0x2000>;
        interrupts = <44>;
        status = "disabled";
    };

    uart5: serial@481a8000 {
        compatible = "ti,omap3-uart";
        ti,hwmods = "uart5";
        clock-frequency = <48000000>;
        reg = <0x481a8000 0x2000>;
        interrupts = <45>;
        status = "disabled";
    };

    uart6: serial@481aa000 {
```

```
        compatible = "ti,omap3-uart";
        ti,hwmods = "uart6";
        clock-frequency = <48000000>;
        reg = <0x481aa000 0x2000>;
        interrupts = <46>;
        status = "disabled";
    };

    i2c0: i2c@44e0b000 {
        compatible = "ti,omap4-i2c";
        #address-cells = <1>;
        #size-cells = <0>;
        ti,hwmods = "i2c1";    /* TODO: Fix hwmod */
        reg = <0x44e0b000 0x1000>;
        interrupts = <70>;
        status = "disabled";
    };

    i2c1: i2c@4802a000 {
        compatible = "ti,omap4-i2c";
        #address-cells = <1>;
        #size-cells = <0>;
        ti,hwmods = "i2c2";    /* TODO: Fix hwmod */
        reg = <0x4802a000 0x1000>;
        interrupts = <71>;
        status = "disabled";
    };

    i2c2: i2c@4819c000 {
        compatible = "ti,omap4-i2c";
        #address-cells = <1>;
        #size-cells = <0>;
        ti,hwmods = "i2c3";    /* TODO: Fix hwmod */
        reg = <0x4819c000 0x1000>;
        interrupts = <30>;
        status = "disabled";
    };

    mmc1: mmc@48060000 {
```

```
        compatible = "ti,omap3-hsmmc";
        ti,hwmods = "mmc1";
        ti,dual-volt;
        ti,needs-special-reset;
        ti,needs-special-hs-handling;
        dmas = <&edma 24
            &edma 25>;
        dma-names = "tx", "rx";
        status = "disabled";
    };

    mmc2: mmc@481d8000 {
        compatible = "ti,omap3-hsmmc";
        ti,hwmods = "mmc2";
        ti,needs-special-reset;
        ti,needs-special-hs-handling;
        dmas = <&edma 2
            &edma 3>;
        dma-names = "tx", "rx";
        status = "disabled";
    };

    mmc3: mmc@47810000 {
        compatible = "ti,omap3-hsmmc";
        ti,hwmods = "mmc3";
        ti,needs-special-reset;
        ti,needs-special-hs-handling;
        status = "disabled";
    };

    wdt2: wdt@44e35000 {
        compatible = "ti,omap3-wdt";
        ti,hwmods = "wd_timer2";
        reg = <0x44e35000 0x1000>;
        interrupts = <91>;
    };

    dcan0: d_can@481cc000 {
        compatible = "bosch,d_can";
```

```
            ti,hwmods = "d_can0";
            reg = <0x481cc000 0x2000>;
            interrupts = <52>;
            status = "disabled";
        };

        dcan1: d_can@481d0000 {
            compatible = "bosch,d_can";
            ti,hwmods = "d_can1";
            reg = <0x481d0000 0x2000>;
            interrupts = <55>;
            status = "disabled";
        };

        timer1: timer@44e31000 {
            compatible = "ti,omap2-timer";
            reg = <0x44e31000 0x400>;
            interrupts = <67>;
            ti,hwmods = "timer1";
            ti,timer-alwon;
        };

        timer2: timer@48040000 {
            compatible = "ti,omap2-timer";
            reg = <0x48040000 0x400>;
            interrupts = <68>;
            ti,hwmods = "timer2";
        };

        timer3: timer@48042000 {
            compatible = "ti,omap2-timer";
            reg = <0x48042000 0x400>;
            interrupts = <69>;
            ti,hwmods = "timer3";
        };

        timer4: timer@48044000 {
            compatible = "ti,omap2-timer";
            reg = <0x48044000 0x400>;
```

```
        interrupts = <92>;
        ti,hwmods = "timer4";
        ti,timer-pwm;
    };

    timer5: timer@48046000 {
        compatible = "ti,omap2-timer";
        reg = <0x48046000 0x400>;
        interrupts = <93>;
        ti,hwmods = "timer5";
        ti,timer-pwm;
    };

    timer6: timer@48048000 {
        compatible = "ti,omap2-timer";
        reg = <0x48048000 0x400>;
        interrupts = <94>;
        ti,hwmods = "timer6";
        ti,timer-pwm;
    };

    timer7: timer@4804a000 {
        compatible = "ti,omap2-timer";
        reg = <0x4804a000 0x400>;
        interrupts = <95>;
        ti,hwmods = "timer7";
        ti,timer-pwm;
    };

    pruss: pruss@4a300000 {
        compatible = "ti,pruss-v2";
        ti,hwmods = "pruss";
        ti,deassert-hard-reset = "pruss", "pruss";
        reg = <0x4a300000 0x080000>;
        ti,pintc-offset = <0x20000>;
        interrupt-parent = <&intc>;
        status = "disabled";
        interrupts = <20 21 22 23 24 25 26 27>;
    };
```

```
rtc@44e3e000 {
    compatible = "ti,da830-rtc";
    reg = <0x44e3e000 0x1000>;
    interrupts = <75
                76>;
    ti,hwmods = "rtc";
};

spi0: spi@48030000 {
    compatible = "ti,omap4-mcspi";
    #address-cells = <1>;
    #size-cells = <0>;
    reg = <0x48030000 0x400>;
    interrupt = <65>;
    ti,spi-num-cs = <2>;
    ti,hwmods = "spi0";
    dmas = <&edma 16
        &edma 17
        &edma 18
        &edma 19>;
    dma-names = "tx0", "rx0", "tx1", "rx1";
    status = "disabled";
};

spi1: spi@481a0000 {
    compatible = "ti,omap4-mcspi";
    #address-cells = <1>;
    #size-cells = <0>;
    reg = <0x481a0000 0x400>;
    interrupt = <125>;
    ti,spi-num-cs = <2>;
    ti,hwmods = "spi1";
    dmas = <&edma 42
        &edma 43
        &edma 44
        &edma 45>;
    dma-names = "tx0", "rx0", "tx1", "rx1";
    status = "disabled";
```

```
};

nop-phy@0 {
    compatible = "nop-xceiv-usb";
};

nop-phy@1 {
    compatible = "nop-xceiv-usb";
};

usb_otg_hs: usb@47400000 {
    compatible = "ti,musb-am33xx";
    reg = <0x47400000 0x1000     /* usbss */
           0x47401000 0x800      /* musb instance 0 */
           0x47401800 0x800>;    /* musb instance 1 */
    interrupts = <17             /* usbss */
                  18             /* musb instance 0 */
                  19>;           /* musb instance 1 */
    multipoint = <1>;
    num-eps = <16>;
    ram-bits = <12>;
    port0-mode = <3>;
    port1-mode = <1>;
    power = <250>;
    ti,hwmods = "usb_otg_hs";
    status = "disabled";
};

mac: ethernet@4a100000 {
    compatible = "ti,cpsw";
    ti,hwmods = "cpgmac0";
    cpdma_channels = <8>;
    ale_entries = <1024>;
    bd_ram_size = <0x2000>;
    no_bd_ram = <0>;
    rx_descs = <64>;
    mac_control = <0x20>;
    slaves = <2>;
```

```
        cpts_active_slave = <0>;
        cpts_clock_mult = <0x80000000>;
        cpts_clock_shift = <29>;
        reg = <0x4a100000 0x800
                0x4a101200 0x100>;
        #address-cells = <1>;
        #size-cells = <1>;
        interrupt-parent = <&intc>;
        /*
         * c0_rx_thresh_pend
         * c0_rx_pend
         * c0_tx_pend
         * c0_misc_pend
         */
        interrupts = <40 41 42 43>;
        ranges;
        disable-napi;

        davinci_mdio: mdio@4a101000 {
            compatible = "ti,davinci_mdio";
            #address-cells = <1>;
            #size-cells = <0>;
            ti,hwmods = "davinci_mdio";
            bus_freq = <1000000>;
            reg = <0x4a101000 0x100>;
        };

        cpsw_emac0: slave@4a100200 {
            /* Filled in by U-Boot */
            mac-address = [ 00 00 00 00 00 00 ];
        };

        cpsw_emac1: slave@4a100300 {
            /* Filled in by U-Boot */
            mac-address = [ 00 00 00 00 00 00 ];
        };
    };

tscadc: tscadc@44e0d000 {
```

```
        compatible = "ti,ti-tscadc";
        reg = <0x44e0d000 0x1000>;
        interrupt-parent = <&intc>;
        interrupts = <16>;
        ti,hwmods = "adc_tsc";
        status = "disabled";
    };

    lcdc: lcdc@4830e000 {
        compatible = "ti,am3352-lcdc", "ti,da830-lcdc";
        reg = <0x4830e000 0x1000>;
        interrupts = <36>;
        status = "disabled";
        ti,hwmods = "lcdc";
    };

    epwmss0: epwmss@48300000 {
        compatible = "ti,am33xx-pwmss";
        reg = <0x48300000 0x10>;
        ti,hwmods = "epwmss0";
        #address-cells = <1>;
        #size-cells = <1>;
        status = "disabled";
        ranges = <0x48300100 0x48300100 0x80    /* ECAP */
                0x48300180 0x48300180 0x80     /* EQEP */
                0x48300200 0x48300200 0x80>; /* EHRPWM */

        ecap0: ecap@48300100 {
            compatible = "ti,am33xx-ecap";
            #pwm-cells = <3>;
            reg = <0x48300100 0x80>;
            ti,hwmods = "ecap0";
            status = "disabled";
        };

        ehrpwm0: ehrpwm@48300200 {
            compatible = "ti,am33xx-ehrpwm";
            #pwm-cells = <3>;
            reg = <0x48300200 0x80>;
```

```
            ti,hwmods = "ehrpwm0";
            status = "disabled";
        };
    };

    epwmss1: epwmss@48302000 {
        compatible = "ti,am33xx-pwmss";
        reg = <0x48302000 0x10>;
        ti,hwmods = "epwmss1";
        #address-cells = <1>;
        #size-cells = <1>;
        status = "disabled";
        ranges = <0x48302100 0x48302100 0x80    /* ECAP */
                0x48302180 0x48302180 0x80    /* EQEP */
                0x48302200 0x48302200 0x80>; /* EHRPWM */

        ecap1: ecap@48302100 {
            compatible = "ti,am33xx-ecap";
            #pwm-cells = <3>;
            reg = <0x48302100 0x80>;
            ti,hwmods = "ecap1";
            status = "disabled";
        };

        ehrpwm1: ehrpwm@48302200 {
            compatible = "ti,am33xx-ehrpwm";
            #pwm-cells = <3>;
            reg = <0x48302200 0x80>;
            ti,hwmods = "ehrpwm1";
            status = "disabled";
        };
    };

    epwmss2: epwmss@48304000 {
        compatible = "ti,am33xx-pwmss";
        reg = <0x48304000 0x10>;
        ti,hwmods = "epwmss2";
        #address-cells = <1>;
        #size-cells = <1>;
```

```
                status = "disabled";
                ranges = <0x48304100 0x48304100 0x80    /* ECAP */
                        0x48304180 0x48304180 0x80    /* EQEP */
                        0x48304200 0x48304200 0x80>; /* EHRPWM */

                ecap2: ecap@48304100 {
                    compatible = "ti,am33xx-ecap";
                    #pwm-cells = <3>;
                    reg = <0x48304100 0x80>;
                    ti,hwmods = "ecap2";
                    status = "disabled";
                };

                ehrpwm2: ehrpwm@48304200 {
                    compatible = "ti,am33xx-ehrpwm";
                    #pwm-cells = <3>;
                    reg = <0x48304200 0x80>;
                    ti,hwmods = "ehrpwm2";
                    status = "disabled";
                };
            };

            sham: sham@53100000 {
                compatible = "ti,omap4-sham";
                ti,hwmods = "sham";
                #address-cells = <1>;
                #size-cells = <0>;
                reg = <0x53100000 0x200>;
                interrupt-parent = <&intc>;
                interrupts = <109>;
                dmas = <&edma 36>;
                dma-names = "rx";
            };

            aes: aes@53500000 {
                compatible = "ti,omap4-aes";
                ti,hwmods = "aes";
                #address-cells = <1>;
                #size-cells = <0>;
```

```
        reg = <0x53500000 0xa0>;
        interrupt-parent = <&intc>;
        interrupts = <102>;
        dmas = <&edma 6
            &edma 5>;
        dma-names = "tx", "rx";
    };

    mcasp0: mcasp@48038000 {
        compatible = "ti,omap2-mcasp-audio";
        #address-cells = <1>;
        #size-cells = <0>;
        ti,hwmods = "mcasp0";
        reg = <0x48038000 0x2000>;
        interrupts = <80 81>;
        status = "disabled";
        asp-chan-q = <2>;    /* EVENTQ_2 */
        tx-dma-offset = <0x46000000>;
        rx-dma-offset = <0x46000000>;
        dmas = <&edma 8
            &edma 9>;
        dma-names = "tx", "rx";
    };

    mcasp1: mcasp@4803C000 {
        compatible = "ti,omap2-mcasp-audio";
        #address-cells = <1>;
        #size-cells = <0>;
        ti,hwmods = "mcasp1";
        reg = <0x4803C000 0x2000>;
        interrupts = <82 83>;
        status = "disabled";
        asp-chan-q = <2>;    /* EVENTQ_2 */
        tx-dma-offset = <0x46400000>;
        rx-dma-offset = <0x46400000>;
        dmas = <&edma 10
            &edma 11>;
        dma-names = "tx", "rx";
    };
```

```
        };
    };


//*********************************************************************
```

D.2   AM335X–BONE–COMMON.DTSI

This section contains the file am335x–bone–common.dtsi which is the devicetree data common for both BeagleBone and BeagleBone Black (includes am33xx.dtsi).

```
//*********************************************************************
/*
 * Copyright (C) 2012 Texas Instruments Incorporated - http://www.ti.com/
 *
 * This program is free software; you can redistribute it and/or modify
 * it under the terms of the GNU General Public License version 2 as
 * published by the Free Software Foundation.
 */

/include/ "am33xx.dtsi"

/ {
    model = "TI AM335x BeagleBone";
    compatible = "ti,am335x-bone", "ti,am33xx";

    cpus {
        cpu@0 {
            cpu0-supply = <&dcdc2_reg>;
        };
    };

    memory {
        device_type = "memory";
        reg = <0x80000000 0x10000000>; /* 256 MB */
    };

    am33xx_pinmux: pinmux@44e10800 {
        pinctrl-names = "default";
        pinctrl-0 = <&userleds_pins>;
```

```
userled_pins: pinmux_userled_pins {
    pinctrl-single,pins = <
        0x54 0x7     /* gpmc_a5.gpio1_21, OUTPUT | MODE7 */
        0x58 0x17    /* gpmc_a6.gpio1_22, OUTPUT_PULLUP | MODE7 */
        0x5c 0x7     /* gpmc_a7.gpio1_23, OUTPUT | MODE7 */
        0x60 0x17    /* gpmc_a8.gpio1_24, OUTPUT_PULLUP | MODE7 */
    >;
};
i2c0_pins: pinmux_i2c0_pins {
    pinctrl-single,pins = <
        0x188 0x70   /* i2c0_sda, SLEWCTRL_SLOW |
                        INPUT_PULLUP | MODE0 */
        0x18c 0x70   /* i2c0_scl, SLEWCTRL_SLOW |
                        INPUT_PULLUP | MODE0 */
    >;
};
i2c2_pins: pinmux_i2c2_pins {
    pinctrl-single,pins = <
        0x178 0x73   /* uart1_ctsn.i2c2_sda, SLEWCTRL_SLOW |
                        INPUT_PULLUP | MODE3 */
        0x17c 0x73   /* uart1_rtsn.i2c2_scl, SLEWCTRL_SLOW |
                        INPUT_PULLUP | MODE3 */
    >;
};
};

ocp: ocp {
uart1: serial@44e09000 {
    status = "okay";
};

gpio-leds {
    compatible = "gpio-leds";
    pinctrl-names = "default";
    pinctrl-0 = <&userled_pins>;

    led0 {
        label = "beaglebone:green:usr0";
        gpios = <&gpio2 21 0>;
```

```
                    linux,default-trigger = "heartbeat";
                    default-state = "off";
                };

                led1 {
                    label = "beaglebone:green:usr1";
                    gpios = <&gpio2 22 0>;
                    linux,default-trigger = "mmc0";
                    default-state = "off";
                };

                led2 {
                    label = "beaglebone:green:usr2";
                    gpios = <&gpio2 23 0>;
                    linux,default-trigger = "cpu0";
                    default-state = "off";
                };

                led3 {
                    label = "beaglebone:green:usr3";
                    gpios = <&gpio2 24 0>;
                    default-state = "off";
                    linux,default-trigger = "mmc1";
                };
            };

            rtc@44e3e000 {
                ti,system-power-controller;
            };
        };

        bone_capemgr {
            compatible = "ti,bone-capemgr";
            status = "okay";

            eeprom = <&baseboard_eeprom>;

            baseboardmaps {
                baseboard_beaglebone: board@0 {
```

```
        board-name = "A335BONE";
        compatible-name = "ti,beaglebone";
    };

    baseboard_beaglebone_black: board@1 {
        board-name = "A335BNLT";
        compatible-name = "ti,beaglebone-black";
    };
};

slots {
    slot@0 {
        eeprom = <&cape_eeprom0>;
    };

    slot@1 {
        eeprom = <&cape_eeprom1>;
    };

    slot@2 {
        eeprom = <&cape_eeprom2>;
    };

    slot@3 {
        eeprom = <&cape_eeprom3>;
    };

    /* Beaglebone black has it soldered on */
     slot@4 {
        ti,cape-override;
        compatible = "ti,beaglebone-black";
        board-name = "Bone-LT-eMMC-2G";
        version = "00A0";
        manufacturer = "Texas Instruments";
        part-number = "BB-BONE-EMMC-2G";
     };

    /* geiger cape version A0 without an EEPROM */
    slot@5 {
```

```
        ti,cape-override;
        compatible = "kernel-command-line", "runtime";
        board-name = "Bone-Geiger";
        version = "00A0";
        manufacturer = "Geiger Inc.";
        part-number = "BB-BONE-GEIGER";
    };

    /* Beaglebone black has it soldered on */
    slot@6 {
        ti,cape-override;
        compatible = "ti,beaglebone-black";
        board-name = "Bone-Black-HDMI";
        version = "00A0";
        manufacturer = "Texas Instruments";
        part-number = "BB-BONELT-HDMI";
    };

    /* Nixie cape version A0 without an EEPROM */
    slot@7 {
        ti,cape-override;
        compatible = "kernel-command-line", "runtime";
        board-name = "Bone-Nixie";
        version = "00A0";
        manufacturer = "Ranostay Industries";
        part-number = "BB-BONE-NIXIE";
    };

    /* adafruit 1.8" TFT prototype cape */
    slot@8 {
        ti,cape-override;
        compatible = "kernel-command-line", "runtime";
        board-name = "Bone-TFT";
        version = "00A0";
        manufacturer = "Adafruit";
        part-number = "BB-BONE-TFT-01";
    };

    /* adafruit RTC DS1307 prototype cape */
```

```
    slot@9 {
        ti,cape-override;
        compatible = "kernel-command-line", "runtime";
        board-name = "Bone-RTC";
        version = "00A0";
        manufacturer = "Adafruit";
        part-number = "BB-BONE-RTC-01";
    };

    slot@10 {
        ti,cape-override;
        compatible = "kernel-command-line", "runtime";
        board-name = "Bone-Hexy";
        version = "00A0";
        manufacturer = "Koen Kooi";
        part-number = "BB-BONE-HEXY-01";
    };
    /* MRF24J40 Cape Override */
    slot@11 {
        ti,cape-override;
        compatible = "kernel-command-line", "runtime";
        board-name = "Bone-MRF24J40";
        version = "00A0";
        manufacturer = "Signal 11 Software";
        part-number = "BB-BONE-MRF24J40";
    };
};

/* mapping between board names and dtb objects */
capemaps {
    /* DVI cape */
    cape@0 {
        /* board-name = "BeagleBone DVI-D CAPE"; */
        part-number = "BB-BONE-DVID-01";
        version@00A0 {
            version = "00A0";
            dtbo = "cape-bone-dvi-00A0.dtbo";
        };
        version@00A1 {
```

```
                version = "00A1", "01";
                dtbo = "cape-bone-dvi-00A1.dtbo";
            };
            version@00A2 {
                version = "00A2", "A2";
                dtbo = "cape-bone-dvi-00A2.dtbo";
            };
            version@00A3 {
                version = "00A3";
                dtbo = "cape-bone-dvi-00A2.dtbo";
            };
        };

        /* beaglebone black emmc on board */
        cape@1 {
            /* board-name = "BeagleBone 2G eMMC1 CAPE"; */
            part-number = "BB-BONE-EMMC-2G";
            version@00A0 {
                version = "00A0";
                dtbo = "cape-bone-2g-emmc1.dtbo";
            };
        };

        /* geiger cape */
        cape@2 {
            part-number = "BB-BONE-GEIGER";
            version@00A0 {
                version = "00A0";
                dtbo = "cape-bone-geiger-00A0.dtbo";
            };
        };

        /* LCD3 cape */
        cape@3 {
            part-number = "BB-BONE-LCD3-01";
            version@00A0 {
                version = "00A0";
                dtbo = "cape-bone-lcd3-00A0.dtbo";
            };
```

```
        version@00A2 {
            version = "00A2";
            dtbo = "cape-bone-lcd3-00A2.dtbo";
        };
    };

    /* Weather cape */
    cape@4 {
        part-number = "BB-BONE-WTHR-01";
        version@00A0 {
            version = "00A0";
            dtbo = "cape-bone-weather-00A0.dtbo";
        };
    };

    /* beaglebone black hdmi on board */
    cape@5 {
        part-number = "BB-BONELT-HDMI";
        version@00A0 {
            version = "00A0";
            dtbo = "cape-boneblack-hdmi-00A0.dtbo";
        };
    };

    /* nixie cape */
    cape@6 {
        part-number = "BB-BONE-NIXIE";
        version@00A0 {
            version = "00A0";
            dtbo = "cape-bone-nixie-00A0.dtbo";
        };
    };
    cape@7 {
        part-number = "BB-BONE-TFT-01";
        version@00A0 {
            version = "00A0";
            dtbo = "cape-bone-adafruit-lcd-00A0.dtbo";
        };
    };
```

```
            cape@8 {
                part-number = "BB-BONE-RTC-01";
                version@00A0 {
                    version = "00A0";
                    dtbo = "cape-bone-adafruit-rtc-00A0.dtbo";
                };
            };

            cape@9 {
                part-number = "BB-BONE-HEXY-01";
                version@00A0 {
                    version = "00A0";
                    dtbo = "cape-bone-hexy-00A0.dtbo";
                };
            };
            /* mrf24j40 cape */
            cape@10 {
                part-number = "BB-BONE-MRF24J40";
                version@00A0 {
                    version = "00A0";
                    dtbo = "cape-bone-mrf24j40-00A0.dtbo";
                };
            };
            /* expansion test */
            cape@11 {
                part-number = "BB-BONE-EXPTEST";
                version@00A0 {
                    version = "00A0";
                    dtbo = "cape-bone-exptest-00A0.dtbo";
                };
            };
        };
    };

    vmmcsd_fixed: fixedregulator@0 {
        compatible = "regulator-fixed";
        regulator-name = "vmmcsd_fixed";
        regulator-min-microvolt = <3300000>;
```

```
            regulator-max-microvolt = <3300000>;
        };

    };

&i2c0 {
        status = "okay";
        clock-frequency = <400000>;
        pinctrl-names = "default";
        pinctrl-0 = <&i2c0_pins>;

        tps: tps@24 {
            reg = <0x24>;
        };

        baseboard_eeprom: baseboard_eeprom@50 {
            compatible = "at,24c256";
            reg = <0x50>;
        };
};

&i2c2 {
        status = "okay";
        pinctrl-names = "default";
        pinctrl-0 = <&i2c2_pins>;

        clock-frequency = <100000>;

        cape_eeprom0: cape_eeprom0@54 {
            compatible = "at,24c256";
            reg = <0x54>;
        };

        cape_eeprom1: cape_eeprom1@55 {
            compatible = "at,24c256";
            reg = <0x55>;
        };

        cape_eeprom2: cape_eeprom2@56 {
```

```
            compatible = "at,24c256";
            reg = <0x56>;
        };

        cape_eeprom3: cape_eeprom3@57 {
            compatible = "at,24c256";
            reg = <0x57>;
        };
    };
};

/include/ "tps65217.dtsi"

&tps {
    ti,pmic-shutdown-controller;

    regulators {
        dcdc1_reg: regulator@0 {
            regulator-always-on;
        };

        dcdc2_reg: regulator@1 {
            /* VDD_MPU voltage limits 0.95V - 1.26V with +/-4% tolerance */
            regulator-name = "vdd_mpu";
            regulator-min-microvolt = <925000>;
            regulator-max-microvolt = <1325000>;
            regulator-boot-on;
            regulator-always-on;
        };

        dcdc3_reg: regulator@2 {
            /* VDD_CORE voltage limits 0.95V - 1.1V with +/-4% tolerance */
            regulator-name = "vdd_core";
            regulator-min-microvolt = <925000>;
            regulator-max-microvolt = <1150000>;
            regulator-boot-on;
            regulator-always-on;
        };

        ldo1_reg: regulator@3 {
```

```
                    regulator-always-on;
            };

            ldo2_reg: regulator@4 {
                regulator-always-on;
            };

            ldo3_reg: regulator@5 {
                regulator-min-microvolt = <1800000>;
                regulator-max-microvolt = <3300000>;
                regulator-always-on;
            };

            ldo4_reg: regulator@6 {
                regulator-always-on;
            };
        };
};

&cpsw_emac0 {
    phy_id = <&davinci_mdio>, <0>;
};

&cpsw_emac1 {
    phy_id = <&davinci_mdio>, <1>;
};

&mmc1 {
    status = "okay";
    vmmc-supply = <&ldo3_reg>;
    ti,vcc-aux-disable-is-sleep;
};

&edma {
    ti,edma-xbar-event-map = <32 12>;
};

&sham {
    status = "okay";
```

```
};

&aes {
    status = "okay";
};

&usb_otg_hs {
    interface_type = <1>;
    power = <250>;
    status = "okay";
};


//****************************************************************
```

D.3   AM335X–BONE.DTS

This section contains the file am335x–bone.dts which is the devicetree data for BeagleBone (includes am335x–bone–common.dtsi).

```
//****************************************************************
/*
 * Copyright (C) 2012 Texas Instruments Incorporated - http://www.ti.com/
 *
 * This program is free software; you can redistribute it and/or modify
 * it under the terms of the GNU General Public License version 2 as
 * published by the Free Software Foundation.
 */
/dts-v1/;

/include/ "am33xx.dtsi"

/include/ "am335x-bone-common.dtsi"


//****************************************************************
```

D.4   AM335X–BONEBLACK.DTS

This section contains the file am335x–boneblack.dts which is the devicetree data for BeagleBone Black (includes am335x–bone–common.dtsi).

```
//****************************************************************
```

```
/*
 * Copyright (C) 2012 Texas Instruments Incorporated - http://www.ti.com/
 *
 * This program is free software; you can redistribute it and/or modify
 * it under the terms of the GNU General Public License version 2 as
 * published by the Free Software Foundation.
 */
/dts-v1/;

/include/ "am33xx.dtsi"

/include/ "am335x-bone-common.dtsi"

&userled_pins {
    pinctrl-single,pins = <
        0x54 0x7    /* gpmc_a5.gpio1_21, OUTPUT | MODE7 */
        0x58 0x17   /* gpmc_a6.gpio1_22, OUTPUT_PULLUP | MODE7 */
        0x5c 0x7    /* gpmc_a7.gpio1_23, OUTPUT | MODE7 */
        0x60 0x17   /* gpmc_a8.gpio1_24, OUTPUT_PULLUP | MODE7 */
        0x00c 0x31  /* P8_6 gpmc_ad3.mmc1_dat1 PIN_INPUT_PULLUP |
                         OMAP_MUX_MODE1 */
        0x008 0x31  /* P8_5 gpmc_ad2.mmc1_dat2 PIN_INPUT_PULLUP |
                         OMAP_MUX_MODE1 */
        0x004 0x31  /* P8_24 gpmc_ad1.mmc1_dat1 PIN_INPUT_PULLUP |
                         OMAP_MUX_MODE1 */
        0x000 0x31  /* P8_25 gpmc_ad0.mmc1_dat0 PIN_INPUT_PULLUP |
                         OMAP_MUX_MODE1 */
        0x084 0x32  /* P8_20 gpmc_csn2.mmc1_cmd OMAP_MUX_MODE2 |
                         AM33XX_PIN_INPUT_PULLUP} */
        0x080 0x32  /* P8_21 gpmc_csn1.immc1_clk OMAP_MUX_MODE2 |
                         AM33XX_PIN_INPUT_PULLUP} */
    >;
};

&ldo3_reg {
    regulator-min-microvolt = <1800000>;
    regulator-max-microvolt = <1800000>;
    regulator-always-on;
};
```

```
&mmc1 {
    vmmc-supply = <&vmmcsd_fixed>;
};

&mmc2 {
    vmmc-supply = <&vmmcsd_fixed>;
    bus-width = <8>;
    ti,non-removable;
    status = "okay";
};


&cpu {
    /*
     * To consider voltage drop between PMIC and SoC,
     * tolerance value is reduced to 2% from 4% and
     * voltage value is increased as a precaution.
     */
    operating-points = <
        /* kHz     uV */
        1000000 1350000
        800000  1300000
        600000  1112000
        300000   969000
    >;
};


//*****************************************************************
```

D.5   AM33XX_PWM–00A0.DTS

This section contains the file am33xx_pwm–00A0.dts which is a devicetree overlay that enables all of the PWMs on the am335x that can be accessed on the BeagleBone headers.

```
//*****************************************************************
/*
 * Copyright (C) 2013 CircuitCo
 * Copyright (C) 2013 Texas Instruments
 *
 * This program is free software; you can redistribute it and/or modify
```

```
 * it under the terms of the GNU General Public License version 2 as
 * published by the Free Software Foundation.
 */
/dts-v1/;
/plugin/;

/ {
    compatible = "ti,beaglebone";

    /* identification */
    part-number = "test1";
    version = "00A0";

    fragment@0 {
        target = <&epwmss0>;
        __overlay__ {
            status = "okay";
        };
    };

    fragment@1 {
        target = <&ehrpwm0>;
        __overlay__ {
            status = "okay";
        };
    };

    fragment@2 {
        target = <&ecap0>;
        __overlay__ {
            status = "okay";
        };
    };

    fragment@3 {
        target = <&epwmss1>;
        __overlay__ {
            status = "okay";
        };
```

```
    };

    fragment@4 {
        target = <&ehrpwm1>;
        __overlay__ {
            status = "okay";
        };
    };

    fragment@5 {
        target = <&epwmss2>;
        __overlay__ {
            status = "okay";
        };
    };

    fragment@6 {
        target = <&ehrpwm2>;
        __overlay__ {
            status = "okay";
        };
    };

    fragment@7 {
        target = <&ecap2>;
        __overlay__ {
            status = "okay";
        };
    };
};


//****************************************************************
```

D.6   BONE_PWM_P8_13–00A0.DTS

This section contains the file bone_pwm_P8_13–00A0.dts which is a devicetree overlay that configures the pinmux and PWM for BeagleBone header pin P8_13. This code example requires that am33xx_pwm be loaded first to enable the PWM subsystems before this overlay configures an individual PWM.

```
//****************************************************************
```

```
/*
 * Copyright (C) 2013 CircuitCo
 * Copyright (C) 2013 Texas Instruments
 *
 * This program is free software; you can redistribute it and/or modify
 * it under the terms of the GNU General Public License version 2 as
 * published by the Free Software Foundation.
 */
/dts-v1/;
/plugin/;

/ {
    compatible = "ti,beaglebone";

    /* identification */
    part-number = "bone_pwm_P8_13";
    version = "00A0";

    fragment@0 {
        target = <&am33xx_pinmux>;
        __overlay__ {
            pwm_P8_13: pinmux_pwm_P8_13_pins {
                /* P8_13 (ZCZ ball T10) | MODE 4 */
                pinctrl-single,pins = <0x024  0x4>;
            };
        };
    };

    fragment@1 {
        target = <&ocp>;
        __overlay__ {
            pwm_test_P8_13 {
                compatible   = "pwm_test";
                pwms         = <&ehrpwm2 1 500000 1>;
                pwm-names    = "PWM_P8_13";
                pinctrl-names = "default";
                pinctrl-0    = <&pwm_P8_13>;
                enabled      = <1>;
                duty         = <0>;
```

```
                        status       = "okay";
                    };
                };
            };
        };

//***********************************************************************
```

D.7   CAPE–BONE–IIO–00A0.DTS

This section contains the file cape–bone–iio–00A0.dts which is a devicetree overlay that utilizes the bone–iio–helper to provide SYSFS entries for the ADCs.

```
//***********************************************************************
/*
 * Copyright (C) 2012 Texas Instruments Incorporated - http://www.ti.com/
 *
 * This program is free software; you can redistribute it and/or modify
 * it under the terms of the GNU General Public License version 2 as
 * published by the Free Software Foundation.
 */
/dts-v1/;
/plugin/;

/ {
    compatible = "ti,beaglebone", "ti,beaglebone-black";

    /* identification */
    part-number = "iio-test";

    fragment@0 {
        target = <&ocp>;
        __overlay__ {
            /* avoid stupid warning */
            #address-cells = <1>;
            #size-cells = <1>;

            tscadc {
                compatible = "ti,ti-tscadc";
                reg = <0x44e0d000 0x1000>;
```

```
            interrupt-parent = <&intc>;
            interrupts = <16>;
            ti,hwmods = "adc_tsc";
            status = "okay";

            adc {
                ti,adc-channels = <8>;
            };
        };

        test_helper: helper {
            compatible = "bone-iio-helper";
            vsense-name  = "AIN0", "AIN1", "AIN2", "AIN3", "AIN4",
                           "AIN5", "AIN6", "AIN7";
            vsense-scale = <100      100      100      100      100
                            100      100      100>;
            status = "okay";
        };
    };
  };
};
//******************************************************************
```

Authors' Biographies

STEVE BARRETT

Steve is a life time teacher. He has taught at a variety of age levels from middle school science enhancement programs through graduate level coursework. He served in the United States Air Force for 20 years and spent approximately half of that time as a faculty member at the United States Air Force Academy. Following military "retirement," he began a second academic career at the University of Wyoming as an assistant professor. He now serves as a Professor of Electrical and Computer Engineering and the Associate Dean for Academic Programs. He is planning on teaching into his 80s and considers himself a student first teacher. Most importantly, he has two "grand beagles," Rory and Romper, fondly referred to as the "girls."

JASON KRIDNER

Jason got an early start with computing at age 9 programming his mom's Tandy Radio Shack TRS-80. He was also a big fan of Forrest Mim's "Getting Started in Electronics." Much of his allowance was spent developing projects. He really enjoyed the adventure of trying new hardware and software projects. His goal is to bring back this spirit of adventure and discovery to the BeagleBoard.org community. While still in high school, he worked extensively with AutoCAD as a leak and flow testing company. He joined Texas Instruments in 1992 after a co-op with them while a student at Texas A&M University. He started using Linux at about the same time. Since joining T.I. he has worked on a wide variety of projects including audio digital signal processing, modems, home theater sound, multi-dimensional audio and MP3 player development.

Index